W9-ACI-746

Storey's Guide to Raising Pigs

STOREY'S GUIDE TO

RAISING
PIGS

· FOURTH EDITION ·

CARE · FACILITIES · MANAGEMENT
BREED SELECTION

Kelly Klober

Storey Publishing

The mission of Storey Publishing is to serve our customers by
publishing practical information that encourages
personal independence in harmony with the environment.

Edited by Deborah Burns and Sarah Guare
Art direction and book design by Jeff Stiefel
Text production by Liseann Karandisecky and Jennifer Jepson Smith
Indexed by Andrea Chesman

Cover photography by © Bloomberg/Getty Images, back; © Jason Houston, front;
 © Uta Ruge/Getty Images, spine
Interior photography by Mars Vilaubi, 68, 72, 163–165; © Russell Graves, ii, viii,
 xiii, 20–22, 27, 31, 33, 35, 314; © Shawn Linehan, 36, 271, 282, 289
Additional photography credits on page 324

Illustrations by © Elara Tanguy, based on original artwork by Elayne Sears and others,
 except for 51, 124, 157, 166, and 176 by Ilona Sherratt, revised from previous editions.

Storey Publishing
210 MASS MoCA Way
North Adams, MA 01247
storey.com

Printed in China by R.R. Donnelley
10 9 8 7 6 5 4 3 2 1

LIBRARY OF CONGRESS CATALOGING-IN-PUBLICATION DATA

Names: Klober, Kelly, 1949– author.
Title: Storey's guide to raising pigs : care, facilities, management,
 breed selection / Kelly Klober.
Other titles: Guide to raising pigs
Description: 4th edition. | North Adams, MA : Storey Publishing,
 [2018] | Includes bibliographical references and index.
Identifiers: LCCN 2018021359 (print) | LCCN 2018027774 (ebook)
 | ISBN 9781635860443 (ebook)
 | ISBN 9781635860429 (pbk. : alk. paper)
 | ISBN 9781635860436 (hardcover : alk. paper)
Subjects: LCSH: Swine.
Classification: LCC SF395 (ebook) | LCC SF395 .K54 2018 (print)
 | DDC 636.4—dc23
LC record available at https://lccn.loc.gov/2018021359

To my wife, Phyllis Klober,

and my grandparents, Kelly and Elizabeth Brewer,

who helped to make all my farming dreams come true.

CONTENTS

PREFACE

Thoughts on the Changing Nature of the Swine Scene in the Last Decade

In the years since the last revision of this book, some remarkable changes have occurred in the way hogs are kept and managed on small farms and holdings and how that pork has gone on to be marketed.

The pork as a commodity trade has gone global, foreign investors have taken up major holdings in the mega-swine corporations, and distrust of the local product and rising income levels have increased the demand for U.S. pork in China and other parts of the Far East. The big Chinese stake in Smithfield may mean even more selling opportunities in the United States for local pork produced in more traditional ways.

While factory-farmed pork is being sold into the global village, Main Street USA has begun expressing a strong demand for more local food items produced in a more traditional and sustainable way. These are the market niches in which big-box store retailing cannot easily enter and compete. It could be called almost artisanal production, though done with a steady eye on the bottom line of the ledger page and with an abiding respect for the animal in question and its true nature.

The small-scale producer today, by doing things the traditional way, is now the outlier, the maverick. This role is one that cannot be co-opted by big packing or factory farming. Theirs is a local product with the daintiest of carbon footprints and, to reach consumers, requires no 1,500-mile trek down the old Monfort Road to the West Coast or from the Carolinas to the Corn Belt and back to consumers in the East.

Theirs is often breed-specific pork, not from sows that have been pushed to wring ever more poundage out of them. It is produced with a transparency and an interaction between producer and consumer that is bringing something of a fine-wine status to hams and chops. It is a type of pork production that would be very familiar to my grandparents.

It has producers working with small numbers of very select animals, a strong reliance on purebred genetics, the use of simple though very distinctive feedstuffs and rations, more seasonal patterns of production, and producer involvement now continuing well beyond the farm gate. It is a template being employed in many areas of animal agriculture, including sheep rearing and the production of an ever more select table egg.

From a herd of just three to five sows, substantial numbers of offspring can be produced even with a seasonal pattern of farrowing. With purebred females producing purebred offspring and a two-litter farrowing calendar, each sow should still produce 16 to 18 young per year. A litter of just 8 grown to a harvested weight of 240 pounds each should then yield more than 1,100 pounds of pork to be marketed directly in any number of ways.

And as purebreds, they could have further value as seedstock, animals to be used for youth project work, feeder stock to be sold to others, or as whole carcasses to be sold into the restaurant trade where tail to snout is the current standard for processing and recipe preparation.

The breeds of primary marketing importance of the moment are the Duroc and Berkshire, due to the recognized eating qualities of the pork that they produce. Research has determined that the Tamworth and Chester White breeds also produce a rather distinctive pork product, and I believe that the Black Poland and Hampshire breeds may do so also. When produced in a more traditional manner and fed for the task, pork has long been known as perhaps the most savory and flavorsome of meat choices.

Putting together a small swine herd, even in the heart of the Corn/Hog Belt, is no longer the simple task it once was. As factory farming gained its grip on hog finishing and pork production, the belief was that the seedstock would continue to come from the independent farm sector. The inverse quickly proved to be true, and replacement females and breeding boars became one more example of in-house production, much like the company newsletter and the formulas for least-cost, computerized rations.

The seedstock trade buckled and a goodly number of once popular swine breeds slipped ever closer to endangered status. To find a needed boar or a couple of foundation females might now require a trip across two or three states even in the Midwest. Fortunately, once established, a purebred livestock herd or flock can become largely self-sustaining, at least producing needed female replacements. Good Hampshires, Black Polands, Spots, and Chester Whites are no longer that easy to find, and there are some real genetic purity issues surrounding some of the largely black and patterned breeds, where it is easier to inject some Pietrain or other breeding to increase muscling levels.

The show-ring and the commercial sector have parted ways in a most dramatic and extreme manner. In many instances, purebred operations have opted almost entirely to the production of show pigs that are most extreme in their patterns of muscling and cover.

Far and away most of the phone calls that I now receive requesting help in locating breeding stock concern the Berkshire breed, with the Duroc moving up as a strong second-position contender. A more distant third is the Hereford, and interest in the Tamworth seems markedly down since the last updating of this book. Interest in the rarer Large English Black and Gloucester Old Spot breeds has changed as it becomes more recognized that these are rather classic mothering breeds with some issues as to finishing and carcass quality. It is the lot of the pig to be pork, and some of them are better at it than others.

Some of the rarer breeds, such as the Guinea Hog and the Ossabaw, have again begun to languish because of issues such as an absence of clear breed type standards, support organizations lacking the structure to promote the breed, and the very real need for breeders to take them up for more practical ends.

At one farm conference I spoke at a couple of years ago, the question was raised as to what exactly does a poorly bred Guinea Hog look like? From the back of the group, a veteran farm woman raised the even more important question of what should a *good* one look like?

The ultimate value of a hog, any hog, will be determined by the carcass that hangs on the rail at the packing plant. And there it will hang in place next to Duroc and Berkshire and other purebred and crossbred carcasses. Why raisers of some of these minor breeds have not gone together to hold type conferences and breeding class shows to resolve questions of type quality, showcase their breeds, increase sales, and further spread breeding animals has long been a real puzzle to me.

This is not to say that the very old breeds aren't worth preserving, but too many raisers have taken them up as if they were show ponies or teacup poodles. Before it ran its course, the fad for Vietnamese Potbellied pigs saw that gene pool watered down in several ways. Breeding in of other swine breeds added new and different colors outside of known breed type. A breeders' market had pigs that should have been neutered sold for outrageous prices as breeding animals. Speculators and naive investors were driven to bankruptcy, and the Potbellied pig itself was left to symbolize all

that is bad with fads in livestock production and the entire exotic animal sector.

The challenge for concerned folks now is to bring back breeds like the Black Poland and Chester—and, yes, even the Hampshire breed needs to be restored to the type and status that it had just 30 years ago. Many believe that we are on the cusp of a real change in swine type, a return to a naturally thicker and more durable hog that produced a far better tasting and cooking pork product.

Pork Production

Demand is steadily growing for pork from hogs produced in rather specific ways.

They can vary from hogs produced on farms where there are no crates to those that are antibiotic-free to those that are fed in very specific ways. All must have the necessary demand and market structure to ensure that they will be cost effective to produce. And some farms are combining production practices to give their output even greater added value.

Antibiotics have long been overused and abused in the production of livestock. There are now regulations in place to limit the addition of antibiotics to livestock rations and require that it be done under a prescription issued by a veterinarian. These regulations are still quite new, with some possible kinks to be worked through, but they are also backed up by what is going on in the marketplace.

A great many fast-food and family restaurant chains have announced that all of their sources of meat and eggs will be antibiotic-free by 2020 or 2025. If you don't believe this is important, note that every time just one chain opts to promote its "riblet" product, butcher hog prices rise by as much as five cents per pound.

Antibiotic-free production will not come with a simple snap of the fingers. At an Amish farm auction a short while back, I viewed a set of hogs in the early stages of an antibiotic-free program, and also in attendance were several younger farmers with similar programs back home. The females varied greatly in size, breed, body type, and general condition due to the difficulty of assembling any number of breeding animals currently outside of contract production. The boars, though representing one of the current hot breeds, were thumping, rheumy-eyed, and showing obvious respiratory stress. The chill of that early March day and the wet snow flurries falling on them didn't help things either.

Sipping from cups of hot chocolate and looking for a barn wall to lean against out of the wind, some of the new-age, antibiotic-free producers began peppering the old guy with questions. Nearly all of their questions had the same answer.

An antibiotic-free hog is not going to just happen. They should be in such production for the long haul and that means learning to breed and feed your way out of a great many different challenges that will be encountered along the way. Most of those producers were starting with swine genetics that were confinement bred and based. They would have to be bred up for their new environment with body type changes as needed, and animals would have to adapt again to a more natural order, There would be setbacks, and it would take time. It was the same lesson the sons of Noah soon had to learn after the door to the ark was opened.

Livestock has to be selected and bred up for new environmental conditions and challenges. An outdoor, antibiotic-free hog is going to need more cover and a deeper and wider chest. Things as seemingly trivial as the diameter of nostril openings will matter. And this year's pig crop should be put together with the intent to breed even better-performing future generations from them.

It took us 10-plus years to get our Duroc sow herd even close to where we wanted it to be as to performance, depth, and consistency in performance and output. When we sought to start a second Duroc line, we drove a lot of high-dollar Duroc females back up the chute to town before we found the one good one on which to found a new line.

And though right for our small farm and markets, she would not have worked everywhere and for every other swine producer. The factory-farm sector has been guilty of trying to make of the hog a creature that nature never intended it to be: lapdog docile for a life in crates and crowded compounds, with a fat rind that's potato-chip thin, bred to the uniformity of a sleeve of saltine crackers, and in any color you want as long as it is white.

Feed Developments

There are some in the alternative pork camp who are guilty of becoming too rigid in their thinking and advocating that the only alternative course of swine production is theirs and theirs alone.

The feed is the fuel for the hog and thus the fuel for the swine-producing venture. Swine rations have seemingly been tweaked and twisted forever. A most major change in my lifetime was the pulling of animal protein sources from swine and other livestock rations. Now there are attempts to create soy- and even corn-free swine rations. Yes, the species began with an animal of the forest floor that would eat pretty much anything it encountered in its wandering, including the odd, slow peasant. And yes, they are still omnivores.

Blame it on where and how I was raised, but the hog and coarse yellow corn grain are one of those truly rare matches that I believe was made in heaven. Without both, there would certainly be no Cincinnati or Chicago.

A hog is most certainly not a ruminant and cannot graze efficiently. At the very best, gestating sows and older growing hogs, animals that weigh more than 125 pounds, might be able to meet 10 percent of their nutrient needs from a good, legume-based browse. Historically, some very good swine rations were formed simply with open-pollinated corn and whey or skimmed milk.

The corn grain might even have been set to soak for a time in the dairy by-product to soften the grain surface and further improve digestibility. The open-pollinated corn of that earlier era had a crude protein content in the 10 to 13 percent range, and some varieties may have even been higher. With simple supplementation and in seasonal systems of production, they could form fairly complete rations for hogs at all stages of development.

There has been a notable reduction in the crude protein content of many modern corn hybrids. Some have been developed for uses other than livestock feeding, including higher carbohydrate content for alcohol production. Forced-air heating processes for drying down corn moisture content following harvest also can reduce its feeding quality.

Corn still remains the most widely used and available and cost-effective energy choice in formulating swine rations. Some pretty exotic alternatives to corn in swine rations have been proposed, but generally at substantially greater costs or resulting in quite complex and more difficult to formulate rations.

Wheat, barley, and grain sorghum or milo can replace corn in swine rations, although best results occur when they replace no more than 50 percent of the corn in a ration. If ground too fine, wheat can cause palatability issues, and milo needs additional supplementation for vitamins A, D, and E. There is a growing amount of non-GMO corn now available for livestock feeding, and some premium markets do exist for hogs fed in such a manner. Added costs to produce a pound of gain must be offset by a higher selling price for the practice to be successful.

Open-pollinated corn may again represent one of the best ration choices for today's smaller-scale, more traditional pork producers. It would certainly add further distinction and appeal to pork from one of the heirloom breeds or pork from hogs that are raised out of doors.

Small-scale producers, whether direct-marketing pork or not, must realize that the greatest number of consumers were, are, and always will be price-driven buyers. And the small farm producer must always be cost-conscious in his or her selection of production practices and inputs.

Those seeking to find ways around the use of soy in swine rations are using all sorts of alternatives from fish product to other bean crops to formulating very complex rations. The more components going into a ration, the more opportunities there are for one or two to be deficient in nutrient content or past their prime and resulting in an improperly balanced blend. Reductions in protein content in a swine ration can impact growth, development, and reproductive performance.

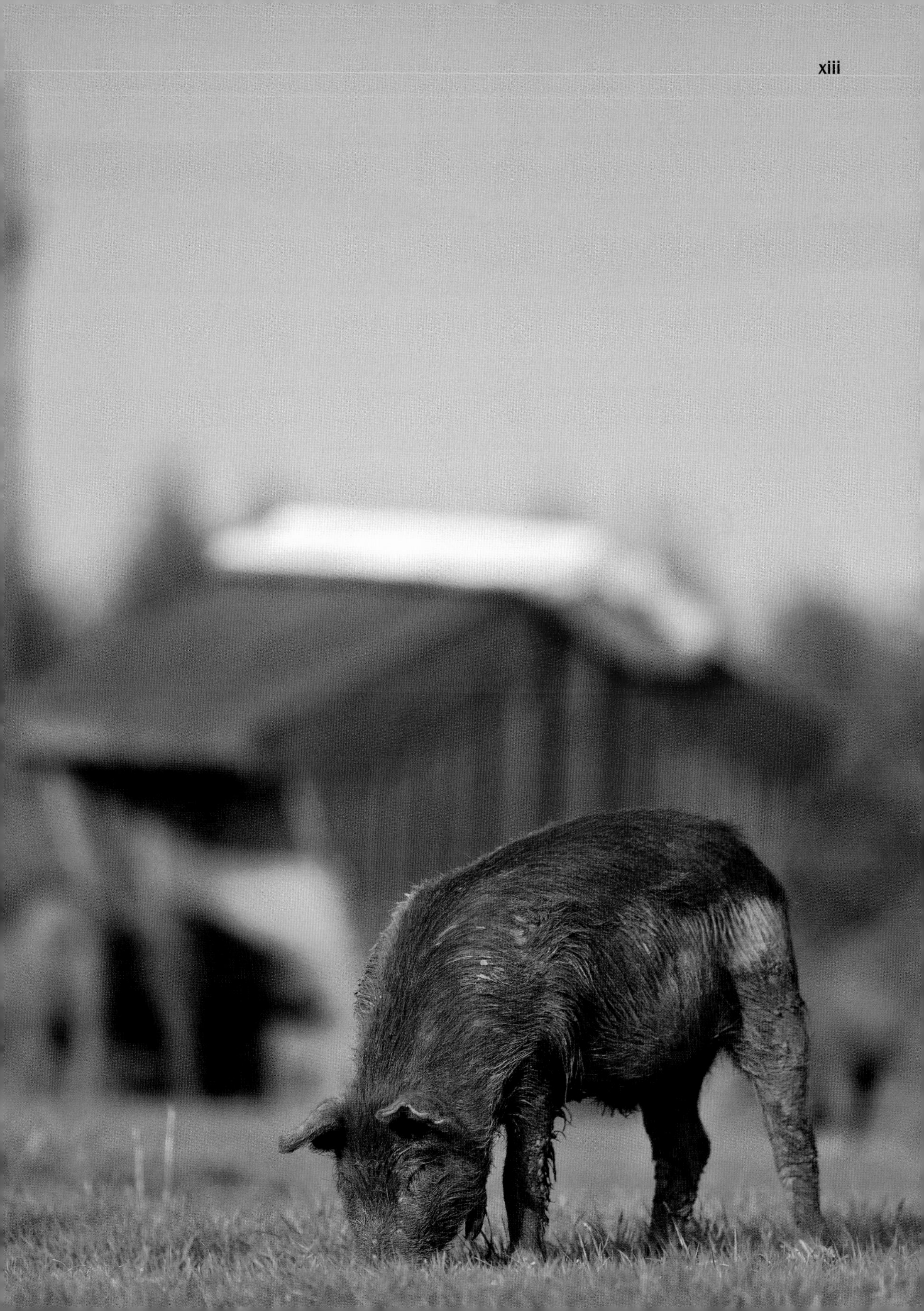

A feral hog can do a fair job of forming its own diet, but with what most consumers would consider an unacceptable "ick" factor and taking a very long time to reach anything near an acceptable harvest weight.

Of late, I have noticed a growing interest in the earlier developed varieties of soybeans in common use when I was a youngster. Then, we were able to save our own soybean seed, and many of those varieties came to fit our farms. Today, they might represent an alternative to the soybean oil meal made from modern soy varieties bred ever more with primary emphasis on yield. Others are using the extrusion process to make a higher-fat soy product from locally grown beans. Most legumes need heat or pressure processing to make them truly useful as a protein source.

Those considering alternative ration components must be assured of selling prices for their hogs or pork that will fully cover all costs. Those costs include possibly added expenses to acquire, necessary on-farm storage, any special processing methods required in their use, and the volatility of price often associated with such less frequently used components. Increasingly, consumers are beginning to question just how much they can afford to pay for value-adding practices, including organic production. The smallholder with limited resources cannot hope to be all things to all people and should focus only on those value-adding practices that will continue to be consistently rewarded in his or her marketing outlets.

Frankly, I have been a bit disappointed by all of the finger pointing that has gone on in the alternative farming camp. Livestock farming can be done in many different ways, and none can be considered to be clearly superior to the others in every aspect. The only people that the producer has to feed are his or her family and self, and to that end the venture must be as profitable as possible. The smallholder who endeavors to operate fairly, to respect the animals in his or her care, and to achieve a good income should be considered a success.

Stock Numbers Reconsidered

Operating with very small numbers is certainly at odds with the thinking that has prevailed in production for the last 50 to 75 years.

Before that, however, it would have been very much in keeping with the way swine production was viewed. A sow herd seldom measured more than 10 head, they were one-boar operations, and they were kept in those numbers for many very good reasons.

They were viewed as one more pursuit among a very rich and diverse mix of ventures that made up the financial undergirding of the family farm. They added to earnings and gave a more even cash flow onto the farm and provided fuller employment for farm family members. Along with the risk management advantages afforded by venture diversity, the added practices gave the farm greater environmental harmony and stability.

Despite all of their claims for extensive biosecurity measures. the massive swine factories are among the most vulnerable of farming institutions. Storm damage, power failure, even terrorist attack—they place tens of thousands of head at near constant risk.

If you've lived any time at all in the country, you know that sheet-metal buildings and chain-link fences, even when topped with razor wire as many are now, are not barriers for mice and sparrows and other forms of vermin that can carry potential health ills with them.

Most vulnerable to health ills are those females giving birth and lactating and the very young. With factory-farming methods, large numbers of both are constantly present on many sites. Some units hold nothing but bred and farrowing females or very young pigs held in hot nurseries. It is a most unpleasant analogy, but their situation is similar to a constant, open sore, vulnerable to all manner of infections.

Rethinking Hog Farming

Today's modern, small-scale producers are beyond outliers; they are charting something rather new, though based on some practices and perceptions of how things should be that are of long standing.

Almost grudgingly, the commercial farm press and some of the elements of the current supporting infrastructure are recognizing that these producers are onto something. It doesn't fit their way of doing things, but it is passing the acid test: it is making money. Even some AI boar studs are starting to carry boars of breeds and type that will fit this more artisanal pork trade.

These pioneering hog men and women are involved in everything from breed preservation work to ration development to direct marketing to end consumers. This is moving producers far beyond when the day would begin by loading some butchers (butcher hogs) on the pickup for sale at the local buying station, visiting a nearby breeder to select a new boar, and then stopping by the local elevator to buy a ton of sow cubes and a couple of hundred pounds of pig starter before heading home.

There are very few markets left in the countryside for selling hogs now, though there does seem to be some rebound in auction sales. Many buying stations may now be open just a

day or two each week or require an appointment to sell hogs.

In nearly every instance now, the new pork producer is creating something rather special and unique for his or her farm and markets. Luckily, the small farm has always been a seat of imagination and creativity.

There are fads and fancies to be worked through. Producers have to reconcile the economics of family farm operation with the wants of foodies and the needs of the great majority of consumers with mortgages and car payments.

Direct marketing now brings producers deeper into the marketing process than at any time since the market fairs of long ago in the Old World. There and then, the local farmers walked the weekly meat needs of nearby villages to town on four good legs.

Direct marketing helps to allay many of the fears consumers have had about the food systems that supply them. It also firmly places the hand of the consumer on the supply tap. What makes it work and what can make it so difficult is that both groups have to come together and create a marketplace face-to-face.

What may matter most in this marketplace is the presence of the producer. What earns a premium is not that the chops are from a Berk, but that they are from the hands of one who knows why that matters, is doing something about it, and is there to relate what has already been done.

This new pork production is certainly at least somewhat familiar to me, but as my dear wife is always quick to point out, I have been going about this for more than 55 years. Recently, however, the governors of North and South Dakota felt it necessary to come together to fund a program to educate younger farmers in basic livestock-raising skills.

The primary markets for coarse grains and soy in this country are still as livestock feeds. The recent return of smaller-scale cattle feeding to the Midwest should also bode well for pork producers as new processing facilities are being built there. The corn and hog bond, I believe, will not be broken.

For the moment, the smaller-scale swine producer is very much an independent agent, largely creating product and market as he or she creates the right swine venture for the home farm. It is reminiscent of my grandparents' day when they went from having a laying flock to operating their own egg route.

They were bypassing middlemen who didn't always care about the farmers' lot and going straight to consumers, who didn't always understand what it meant either and what had to be done to be a successful farmer. It is the producer who is giving renewed life and definition to the concepts of heritage and local pork.

Market Directions

Farmers' markets and CSAs, heritage swine breeds, artisanal and local pork, tail-to-snout eating, and new pork products and processing methods: it is not the hog farming business that I grew up in.

It is not that different, however, and the new pork is joining the cage-free heirloom egg on the breakfast plate of modern and more informed consumers. It is put there by consumer wants and by swine producers determined to do their farming in a better and more natural way.

This is the twenty-first century, but many of the core measures and aspects of swine care remain the same, and many of those practices

and all of the swine breeds still in place were known in the nineteenth century. There are some fewer breeds and perhaps fewer producers, although at the darkest of times in rural America a good many still kept a few sows tucked away behind the machine shed. The local butcher and the small-scale packers have been hard pushed by the big-box store system of food retailing, but they, too, are staging something of a comeback.

Elite chefs are learning to do their own butchering, and there is a growing awareness that pork, like wine, can reflect a certain terroir. In its flavor and texture are vibrant indications of how it was bred and reared and fed.

A couple of times in my life I thought we got things just about right in the hog game, and both times, the independent producer was the one in charge of swine type and production practices.

I was still in my teens when the market was served and very content with a 220-pound butcher hog. They were quite young at harvest, coming off the farm at 5 to 5½ months of age. As young animals they were naturally leaner, were quite efficient in the conversion of feedstuffs to muscle gain, and were producing the smaller primal cuts in keeping with the already developing trend to smaller families and more meals eaten away from home.

The lighter-weight butcher hog was nature's route to a lean and most efficient-to-produce pork product. It was easy to produce, handle, and house.

The packing trade wanted bigger carcasses to get more product from the killing floor and packing line, where there are wages to be paid. The quicker the butcher hog became a mini-beeve, the better. And the push continues for a 300-pound (if not bigger) animal at harvest.

To get them bigger means that they will be older, they will stand around on the farm longer, and they will eat more. The older the animal, the slower it grows, less lean muscle is produced, and the fat level on the carcass increases. A pound of fat requires more feed to produce than a pound of lean gain. Some of these hogs are now on the farm for 180 days or more and sold through a wholesale marketing system where the only premium is for poundage, on and off the killing floor.

In the early 1990s, we again got close to a pretty good hog for the producer and consumer alike. We had a 240-pound butcher animal carrying a backfat cover in the 1-inch range. It was built on a strong purebred base, often derived from a three-breed cross to maximize hybrid vigor in a way still manageable on a small farm. It was a hog that would work in the dirt or on concrete, and a handful of gilts could still get you into farming.

We had a Duroc sow herd with a 10-pig litter average under the shade of white oak trees, and friends and neighbors with herds of Hamps and Yorks and Chesters all doing well, too.

The solutions to many of today's challenges in swine production were largely in hand in earlier times. To those with humane concerns, those lighter-weight market animals held merit. For example, with younger animals at harvest the need for castration might very well be eliminated. And the seedstock producing those later, 240-pound butchers took steamy August days and cold January nights equally in stride.

As this new pork and the new pork producers are being created, they have a past from which to draw. Their future will involve both innovation and renewal, and there should be room for a great many more of them if they remain respectful of the animal and the better traditions behind it.

1

AN INTRODUCTION TO RAISING HOGS

O f all of the major livestock species, none is more misunderstood or less appreciated than the hog. Research now shows that the hog was domesticated at least 6,000 years ago, making it one of the livestock animals with the longest association with man. Both species, I believe, are improved by that association.

The rooting, squealing, twisted-tailed "mortgage lifter" of the Midwest is actually known and valued far beyond the borders of the Corn Belt and the rim of the breakfast plate. The ancient Egyptians sowed seeds into ground broken and trampled by the sharply pointed hooves of hogs, and that practice continues down to this very day with smallholders — rural-lifestyle folks and small farmers who regularly use a hog or two to till their fallow garden sites.

Hogs arrived in the New World with the conquistadors, brought as a trail-along larder, and moved ever westward with the expanding frontier. Along the American frontier, cured pork products — such as hams and bacons — even served as an early form of currency.

In the classic folk song "Sweet Betsy from Pike," a spotted hog figures quite prominently in the livestock inventory that Betsy and her husband, Ike, take west with them. The Pike in that song title was a reference to Pike County, Missouri, which to this day remains one of the best-known pork-producing counties in the nation and is still home to some good Spotted hogs. The first Duroc gilt I bought, now more than 50 years ago, hailed from that same Pike County.

The vigor and hardy nature of those early ridge runners and nearly feral hogs continues. No major livestock species is more durable, adaptable, or productive than the hog. Here on our smallholding in eastern Missouri, the River Hills region, I have had sows produce three litters in a single 12-month period. An average of 2.25 litters per sow per year, with seven or eight pigs weaned per litter, is doable on even the smallest of farms. There are sows on medium-sized farms in both Europe and the United States that are producing nearly 26 pigs each per year. Corporate farms boast an average of more than eight pigs per litter produced by sows bred for confinement, but they are not nearly as quick to note the near 20 percent death loss encountered in the span of time between birth and weaning.

Shoats showing strong Yorkshire breeding resting atop a good straw bed

Small producers with simple facilities following a more seasonal plan of production can realize just two litters per sow per year without taxing their time or resources. And it is reasonable to expect those litters to have an average of eight or nine pigs at both birth and weaning.

Hog-Farming Trends

When Europeans first began to settle in the United States, feral animals were simply harvested as needed from the wild. Later, when America's first farmers began raising pigs, production was one of a diverse mix of modest yet complementary agricultural ventures. This method of production provided the fullest possible employment for the farmer and his family and afforded them a livelihood that was both environmentally and economically sound.

In my early years in the 1960s, hogs were thought to be everybody's "poor relations," valued by farmers only for their year-round marketing potential, not because they were glamorous in any way. Between 1968 and 1985, they became the gateway venture into the farming life. In 1997, when *A Guide to Raising Pigs* was first published, pork production in the United States was entering a time of great change. The break between large and small producers, though long predicted, came about quite suddenly. Hogs were considered an "übercommodity" — the follow-up species to the confinement-housed broiler chicken and the focus of a high-volume, intensely mechanized system.

There followed a period of great realignment and uncertainty when most of the swine producers with medium-sized herds left the field. Production costs increased, while the fees paid for butchering hogs fluctuated widely. Corporate sow herds came to number in the tens of thousands as marketing outlets for small producers declined sharply in numbers. Some swine breeds sank ever closer to extinction, and environmental and social issues gathered over store meat counters like storm clouds. It became harder and harder to actually see hogs nosing about even here in the heart of the Corn Belt. Corporate hogs were concealed in their factory-farm buildings, and small producers numbered a scant few.

In order to survive, small hog producers found they needed to return to older methods and patterns of production — those practiced on the country's early farms. They needed to practice diverse farming methods that were unfamiliar to more recent generations of pork producers. The golden age of the "hog man," the producer specializing solely in pork production from his or her family-farm base, had come to an end. And it was just as well; the single-commodity production agriculture that had evolved was neither good risk management nor consistent with the nature of family farms.

Moving Away from "Big Pork"

There are a number of truly monstrous swine-production facilities operating now, and the general belief has been that it takes a yearly production of at least 50,000 butcher hogs for someone to be considered a hog man in this new age. But the "bigger is better" trend is waning — and it is alive only in the corporate sector. The presence of this sector is not as enduring and as unchangeable as many once believed it to be. "Big pork," for all its millions of pounds of production, is in a precarious position of sorts. It has enemies, a near legion of them.

And the large producers seem to be ever at war with each other, swallowing each other

TERMS USED FOR HOGS AND THE STAGES OF SWINE DEVELOPMENT

GENERIC TERMS

hog: a swine that weighs more than 120 pounds

pig: a male or female swine weighing up to 60 pounds

piglet: layman's term for a young pig, but don't use this around swine producers unless you want to be laughed out of the coffee shop

shoat: a pig, male or female, from 60 to 120 pounds

swine: the species

MARKET TERMS

butcher: a hog being raised to readiness for slaughter

feeder pig: a pig weighing between 35 and 70 pounds and purchased to be fed out (raised) by another swine producer

growing/finishing hog: any hog weighing more than 50 pounds in the feedyard

TERMS FOR FEMALES

dam: female parent

gilt: a young female

open gilt: a female that has reached puberty but has not yet been bred

second-litter gilt: a marketing term; a young sow

sow: a female hog after one or two litters

TERMS FOR MALES

barrow: castrated male

boar: male with testicles intact

sire: male parent

stag: older male; older castrate being sent to slaughter

---

up in mergers at every turn. In their quest for a uniform end product, they embrace technology and practices that leave them with hogs with an ever-narrowing gene pool. It's as if they believe that they somehow have the right to impose their corporate will on the countryside and their consuming public.

There is a strong belief in the traditional rural community that big pork may eventually depart these shores for locales with labor, environmental, and animal-welfare regulations that are friendlier to their corporate farming model. In the manner of today's high-volume fishing fleets, there is speculation that they could load hogs raised anywhere in the world aboard large, seagoing abattoirs and then deposit finished pork products on our shores or wherever the demand is highest. What they will do with the viscera is anyone's guess, but I sure wouldn't want to swim in the wake of any of those ships.

Corporate farms have endured because they have set the bar for bargain-seeking consumers and are willing to slug it out with each other for a penny or two of profit per pound of production. How well that market will continue to sustain producers in the future is debatable now, for just as the hog-producing sector broke in two, so did the consuming public. As American consumers grow ever more selective about their purchases, basing their buying decisions on practices used to raise the food they eat, it can only bode well for smaller hog raisers whose approach to pork production is respectful of consumers' desires and the pigs.

Not long ago, I sorted through a large display of frozen sticks of bulk pork sausage at a local supermarket. They were flavored and spiced in a number of varieties, had all sorts of brand names that used the words "farm" and "country," and used numerous barns and other rural images in their logos, yet *nearly all of them came from just two packing firms with extensive sow herds kept in confinement.*

To some these giants may appear too big to tackle, but to me they actually seem to be more dinosaur than rocket ship to the future. They have grown too big and ponderous. They spend a lot of time combating dissatisfied consumers, rural neighbors, and air-pollution regulations. Pressure to leave the stage is mounting, and as they begin to falter, choices for consumers are growing as a number of marketing and production options open for pig raisers with the creativity and flexibility long identified with America's small, family farmers.

Getting the Whole Hog Picture

It seems that anyone with an interest in a sustainable system of pork production will need to endure a period of transition, perhaps great transition. The independent producer will be pursuing a very different production model from those of the recent past but one not all that different from the way hogs were kept and pigs were produced in the United States in earlier years. Production will not be about pork as a mere commodity. A pig is far more than input loaded into a confinement unit with pressure applied in the form of feed and energy until it is complete and the wastes roll out from one end and the often restructured "pork product" from the other.

More and more, American consumers — with some of the highest disposable incomes in the world — are making informed decisions about food-item and food-supply purchases that are in keeping with their values. They have concerns about the condition of the environment, the quality of life of livestock, the sustainability of freestanding business, and the local agricultural economy. They want smaller, family farmers to be *their* farmers for the future and for the here and now. Hog producers must painstakingly approach the job with all the heart it takes to create fine wine, a rare and sumptuous fruit, or a gourmand-inspired heritage broiler. They will be a part of what is coming to be known as artisanal agriculture.

Successful Small Producers

Emerging now is a type of pork production that is truly imaginative, hinging greatly on **direct marketing**. Small producers are working with quite modest numbers; they are often one-boar operations with seldom more than 10 or 15 sows. My grandfather would recognize and be quite comfortable with this model of production. And the meat produced in the process suits my grandmother's classic Midwestern/Southern style of cooking, as well as that of today's top chefs.

These small businesses focus on artisanal production and consider consumer values the foundation for their business and raising practices. Most of the marketing entails making contact directly with the consumer, rather than relying on retailers and contract processors to do the work. And communication with the customer doesn't end once an order is placed; the producer maintains contact with the consumer during production. These operations are clear alternatives to corporate production.

Some decry this production process as "boutique farming," but call it what detractors may, these hog raisers *are being rewarded with boutique earnings!* Although hog production has been forced into a corporate mold in recent years, those practices don't consider the nature of the hog, the laws of the marketplace, the wishes of the consumer, and the calling of the farmer to be a wise and thoughtful steward. Production on a modest scale, such as that described in this book, is production on a human and humane scale. It requires a great deal of interaction with pigs as well as with the consumer. Although they have sometimes been temporarily overrun by greedy corporate business-management systems, farming operations that keep consumer concerns in the forefront have always been rewarded in the U.S. marketplace.

The versatile hog will always be a creature of the small farm and the smallholding; it is a far too valuable and utilitarian creature to think it might serve otherwise. Its role is not so much changing as it is returning to that of earlier times. You cannot discuss sustainable agriculture without factoring in the importance of extensive venture diversification on family farms. It is a way of life that requires the farmer to engage in 5 to 10 different cost-efficient, easy-to-start ventures that do not compete for farm resources, and offer year-round selling options. I can think of no other large animal

that matches the production and marketing potential of the hog.

To those who are new to raising pigs or who are considering making pig raising a vocation instead of a hobby, I say, bring an open mind to these animals. Study their past very carefully: you will find in the history of hog raising many clues to the future. Stay open to the widest of possible futures. Hog production now will very much be what you, the producer, make of it.

For a sample of a small-producer calendar year with farm ventures that do not compete, see the appendix.

Modern Pork Production

The pigs of today are efficient converters of feedstuffs to lean, healthful protein that is nutrient dense and readily digestible. On average, the pork produced today is a full 30 percent leaner than the pork of the 1950s, when much of the human nutritional data that is still in use was gathered. Through selective

Most hogs as we know them today are thought to have descended from the Eurasian wild boar. From Asia, they spread to Europe and Africa.

breeding, better care and management, and improved feeding practices, modern pork has become a source of protein of the highest quality. It is rich in such high-value nutrients as iron and zinc and has a fat content comparable to a same-size serving portion of chicken.

Lean or Lard?

The modern hog is a meat animal of truly extraordinary abilities. In fact, a distinction is no longer made between bacon- or meat-type hogs and lard hogs. Only once in my lifetime have I seen a drove of any size of No. 4 (too fat, too finished) market hogs, and that was

SERVING SIZES AND NUTRITIONAL PROFILES OF LEAN MEATS

3 ounces of cooked, roasted, trimmed	Calories	Total Fat (g)	Saturated Fat (g)	Cholesterol (mg)
Lean Chicken				
Skinless chicken breast	140	3.1	0.9	73
Skinless chicken leg	162	7.1	2.0	80
Skinless chicken thigh	178	9.3	2.6	81
Lean Cuts of Pork				
Pork tenderloin*	139	4.1	1.4	67
Pork boneless sirloin chop**	164	5.7	1.9	78
Pork boneless loin roast*	165	6.1	2.2	66
Pork boneless top loin chop**	173	6.6	2.3	68
Pork loin chop**	171	6.9	2.5	70
Pork boneless sirloin roast*	168	7.0	2.5	73
Pork rib chop**	186	8.3	2.9	69
Pork boneless rib roast*	182	8.6	3.0	70
Lean Cuts of Beef				
Beef eye of round*	141	4.0	1.5	59
Beef top round***	169	4.3	1.5	76
Beef tip round*	149	5.6	1.8	69
Beef top sirloin**	162	5.8	2.2	76
Beef top loin**	168	7.1	2.7	65
Beef tenderloin**	175	8.1	3.0	71
Fish				
Cod†	89	0.7	0.1	40
Flounder†	99	1.3	0.3	58
Halibut†	119	2.5	0.4	35
Orange roughy†	75	0.8	0.0	22
Salmon†	175	11.0	2.1	54
Shrimp††	84	0.9	0.2	166

*roasted; **broiled; ***braised; †dry heat; ††moist heat
National Pork Producers Council

more than 35 years ago. Even breeds and lines strongest in genetic traits for litter size and mothering ability rather than leanness produce trim carcasses of exceptional merit.

There are now boars approaching that fabled 2:1 feed-efficiency ratio — that is, producing 1 pound of gain on 2 pounds of feed-stuff — a ratio at which pork rivals poultry in production costs. Swine carcasses are getting longer, with a great many now exceeding 33 inches in length; **loineyes** (meat of the loins) are increasing to well past 6 square inches, and plate-sized pork chops are possible through bigger loins, not smaller plates. Many market hogs now exceed 70 percent in the amount of lean they yield when dressed; very little is lost in the way of fat trim. A few years back, we had a gilt processed that was so lean we had to buy fat to add to her trim to make pork sausage with desired cooking qualities.

Today's Pork and Porklike Products

Today's pork comes in the same popular forms as it did years ago. Bacon, ham, sausage, and Canadian bacon still have a solid claim on that first and most important meal of the day: breakfast. The ham sandwich in all sorts of variations remains one of the leading sellers in the restaurant trade, and newer pork products such as the pork burger have had a favorable reception. Baked ham and pork chops are both ranked among the classic comfort foods. Pork is the good stuff, and as the chart on page 6 shows, in all of its forms, it has a role to play in providing a healthful diet.

Today's informed consumers still have a taste for good pork and appreciate its rich and savory nature. However, they often find modern pork produced in confinement lacking in these traits. Pale, soft, and watery, confinement-produced pork became an early exemplar of Frankenfood — unnatural foodstuff named for the creator of the lab-concocted monster from Mary Shelley's 1818 novel. Recent trends in the raising of uber show pigs with extreme loins and hams didn't help to deter that image.

Raisers and consumers are gradually coming to appreciate hogs with a more practical, natural, middle-of-the-road body type and are finding that these more-classic traits are prized for the stove and plate. Those with hogs producing the delicious pork of my childhood memory are finding some highly rewarding market niches now.

The Pork Producer

The geographic positioning of pork production is changing rapidly now, and I am getting the phone calls to prove it. I have fielded calls from swine raisers who are just beginning, as well as those who are expanding their ventures in central Florida, Nevada, Arizona, upstate New York, Texas, Nebraska, Tennessee, and Alabama. This comes as no surprise, however, as there is a pork-producing tradition in nearly every region of the United States. New England was once known for its maple- and applewood-cured pork products and the Eastern Seaboard for classic Smithfield hams (from the famed curing region of Virginia, not the product of the megahog corporation). Every county in the South seems to have its own barbecue sauce recipe; in the Midwest, summer means pork steaks; and pork is a staple of Southwestern and Hispanic cooking. The use of pork in the cuisine of so many diverse cultures and ethnicities signifies the possibility of growing opportunities for pork production in the western United States and in many other areas of North America.

These new producers, along with those who are established but changing their production methods, are operating with a playbook of just a few pages but one with all sorts of room for creativity and imagination. They find support for their methods in both the temper of the times and the historical roots of pork production within the United States.

The basics of the new business plan include:

- Emphasizing the importance of meat quality and the life of the animal over quantity of meat produced
- Direct marketing: One-to-one relationships and becoming "the farmers" for "their patrons," now crucial to the success of the business
- Increasing use of purebred livestock for greater marketing options and personal control over the most important of all production investments, the **seedstock**
- Eagerly embracing the concepts of eating and marketing locally
- Working with modest numbers, often with sow herds of just 5 to 10 head; these are, essentially, one-boar herds, and when kept to these numbers, they pose no potential risk to the environment and will not skew the farm's course so that hog production becomes too great an overall focus
- A commitment to sticking with pork production for the long haul, which means planning and managing accordingly

Many people believe they must have a large acreage and hogs by the hundreds for pig

> ### SPACE REQUIREMENTS
>
> Space for the pig or two you're feeding out for your table can often be measured in mere square feet. One-quarter acre of pasture can support up to four nursing sows and their litters. Few are the smallholdings that can't support a small swine venture of one sort or another.
>
> On our 2.86 acres we regularly ran 7 to 10 sows, produced 25 to 40 breeding-age boars each year, raised our female herd replacement, fed some market hogs, and marketed the rest of the pig crop as **feeders** — feeder pigs that we sold to other producers who fed them to market weight.

production to be a viable enterprise. Nothing could be further from the truth. We once sold two feeder pigs to a couple who asked if their 20 acres of pasture would be large enough to accommodate their new purchases. On that 20 acres, there was certainly room for those two shoats — along with hundreds and hundreds more if housed and contained properly. On our own 2.86 acres we have worked with as many as 10 sows while also managing several other small livestock and poultry ventures.

Hog Myths

Since I've started myth busting, let me address several other mistaken beliefs about hogs and hog raising:

Hogs are inherently dirty animals. This just isn't true. Hogs do lack the ability to sweat and sometimes run into problems in very hot and humid weather because they can release heat only through panting and the evaporative cooling process. They wallow in muddy water to cool off but otherwise have no real affinity for mud. In small lots and around feeding and watering equipment, their sharply pointed hooves tear the soil surface and can cause muddy areas to develop.

Their rooting activity in the quest for mast (roots, fallen nuts and fruits, grubs, and other natural feedstuffs found at or just below the soil's surface) also muddies their image. This is an instinctive feeding process passed down by their feral ancestors, which were denizens of the forest floors of Europe. A hog on pasture or in a wooded lot can still do much to balance its own diet. Although it is no longer an absolute necessity to provide pasture and browsing land for hogs, I believe there is no replacement for the sun shining down on a hog's back and the wind blowing across it. This is nature's way. The hog gets natural vitamin D, and the pork improves naturally with correct fat cover and improved muscle tone due to more exercise.

Hogs are inherently greedy or gluttonous creatures. Hogs are mistakenly assumed to be gluttonous because growing hogs and sows nursing young are fed to full appetite (fed to consume roughly 3 percent of their bodyweight daily) to foster rapid growth and heavy milk production. It is also part of their herd instinct for all the hogs in a group to rise up and want to eat at the same time. Furthermore, most sows are fed limited quantities throughout much of their gestation period; if she amasses less fat, the sow carries the litter more easily and the delivery is less stressful. This limited feeding explains why sows are eager to eat when feed is presented to them.

Hogs are very efficient users of feedstuffs, often averaging 1 pound of gain on just 3 to 3.5 pounds of feedstuffs, even in the simplest of facilities. While this may not be a trait that first-world weight-conscious humans would find becoming in their own physical development, hog producers gratefully accept the fact.

Hogs are accustomed to rooting about for at least a portion of their diet, but they are not eaters of swill and garbage as is so often depicted. In earlier days, they would root through the streets and alleys of rural villages, but their quest was for table scraps, mill wastes, and spilled grain.

The base ingredient in most modern swine rations is one of the coarse feed grains, such as corn, grain sorghum, or barley, and the most commonly used protein supplement is soybean oil meal. The corn is not the sweet corn we humans enjoy but yellow field corn, which is quite low in **crude protein** content. When in high school, I was told to assign a crude protein value of 9 percent to yellow corn when formulating swine rations. Today's corn hybrids seem to be selectively bred for everything but protein content, and when such corn is then run through a dryer, that protein-content figure can suffer even more. Have your grain supply tested for feeding quality, or err on the side of reduced protein content. Some have found that their corn tests down to a crude-protein level of just 6 percent.

Those who think that hogs are crude eaters should bear in mind that they have a digestive system quite similar to our own and are often used in human-related medical research. Other than humans, hogs are the only creatures that will consistently and willingly consume alcohol for study purposes, according to research conducted by the University of Missouri.

You have to raise and sell swine in huge numbers to make money. There are few livestock species that present as many marketing opportunities as hogs: butcher stock, feeder pigs, show pigs, crossbred and purebred breeding stock, roasting pigs, the direct sale of meat animals and pork products, and more, many more. Pork sausage from simply reared hogs easily brings $3.50 a pound; young boars, while in less demand now, still bring up to $500; and there is a modest but quite active small-lot market for young feeder pigs of 70 pounds or less to be fed out for harvest. The key to financial success depends on whether you are practicing good marketing or merely dumping your output on whatever market outlet is available at the moment. But don't get greedy. Although the niche market is currently wide open and expanding, many markets are capable of absorbing only modest amounts at truly premium prices. Some may be supplied by just one farm.

Hogs are more vicious or treacherous to work with than other livestock species. Granted, one of our 7-pound Buff Orpington hens is easier to handle than a 500-pound sow on a one-to-one basis, but I bear a number of chicken-related scars and none from working with hogs.

With hogs, you simply have to build housing and handling facilities that are sturdy enough for such large animals. Housing for hogs should be reinforced at the ground and at hog height, where the animals are, and not at head and shoulder height where we humans most experience our environment. Furthermore, anyone working with hogs should respect them for their strength and size and their often quite surprising speed and agility.

The tales of mean hogs date back to the times when they were truly free range, when you were as likely to encounter feral hogs as domestic ones. If you were attacked, the standard advice was to tuck in your head, cover it with your arms, and draw up your legs to your body. We worry about this far less today; for a great many generations now, hogs have been bred selectively for a more docile nature and quiet temperament.

See chapter 3, page 53 for a discussion of humane handling of hogs.

Talkin' Hog

Visit even briefly the sale-barn alleys, show barns, or hog lots, and you will find the people there speaking a language uniquely their own. Not only will there be talk of gilts and barrows ("bars" in parts of the South and Midwest), but you will also hear about pigs with "daylight" that are "blown apart" or "coon footed."

Each new trend in swine type seems to add at least half a dozen new words to the swine raiser's vocabulary. Still, this is also a language with ancient roots. Where else would you find ancient words like *farrow* and *sow* used in the same sentence with *sonoray*?

What really separates the swine pros from the tenderfeet is how the word **pig** is used. To be country correct, it is the term for a very young pig. A **hog** is a swine that weighs more than 120 pounds. One farmer might tell another that the 400-pound boar standing before them is a "pretty good ol' pig," but the greenhorn who then strolls up and says, "Wow, look at that big pig" has shown himself to be totally green.

Farrow means to give birth. A **barrow** is a castrate, and a **gilt** is a female younger than 12 months of age that has yet to give birth. If she's pregnant, she's called a **bred gilt**. A **shoat** is a recently weaned pig. A **boar** is an intact male, and a **sow** is any female that has had a **litter** of pigs. **Stag** is a now little-used term for an older male to be sold for slaughter.

Older males and females retain good value in the slaughter trade for use in the manufacture of sausage and other highly spiced meat products. **Whole-hog sausage** was once generally made with the pork from sows. However, with the 20-year trend toward production in factory farms and animals kept inside **confinement** buildings with controlled environments, the heavier, higher-yielding sows are in limited supply, and most confinement-held sows are culled by or before their fourth **parity** (fourth time giving birth) and at a weight of 400 pounds or less. Heavy sows — 550 pounds and more — now often bring as much per pound as *No. 1* butcher hogs, sometimes more. Confinement-bred sows stay small, burn out at an early age, and are often sold in a thin and ragged state.

A **butcher** or **market hog** is one weighing 220 to 260 pounds and ready for sale for slaughter. These generally are 5 to 7 months of age. A **feeder pig** is an animal typically weighing 40 to 70 pounds that is sold to a farmer/feeder to be **fed out** (raised and fed) to market weight. Such pigs are generally between 8 and 12 weeks of age. They can be a bit heavier or lighter in weight. A **finishing** hog is a hog

A feeder pig generally weighs 40 to 70 pounds and is between 8 and 12 weeks of age.

that weighs between 100 pounds and market weight.

Every change in swine **type** (body and growth type) — and by some accounts, there may have been as many as 20 in the twentieth century alone — produces several new descriptive terms such as those noted above. Short, thickly made hogs common early in this century were called **cob rollers** because their bellies would roll across the corn cobs left for feed in the pen or lot; later, taller and flatter-muscled animals with long legs were called **race horses**. An animal that is **blown apart** demonstrates good internal body dimension throughout its entire length. In other words, there is visual indication of body capacity and internal-organ development. A good **internal body dimension** is associated with hardiness and more efficient growth. Efficient growth, or **feed efficiency**, refers to how much feed it takes for growth: Good feed efficiency means it takes less feed to achieve good growth.

A pig with **daylight** has good leg length, because you can see some daylight underneath the animal. A **coon-footed** hog walks with a flatter and more flexible foot and has sloping pasterns that indicate it should stand up well on concrete or other unyielding surfaces. It has a "foot" that is, in many ways, akin to the foot of a raccoon.

Throughout this volume, I will use many of the terms in the modern swine raiser's lexicon. Some of them have been around for centuries, others for decades; some come and go in a year or two; and one or two always seem to be newly working their way into the swine raiser's vocabulary. A few years ago the terms **spongy** and **Jell-O-middled** hogs came and went in popularity in little more than a matter of months. Such hogs had a thick middle

when young; many believed this translated into a wider, thicker meat hog. Sometimes it did, but sometimes it resulted in a hog that was just fat. On the other hand, your great-grandfather would find assurance in knowing that a hog with a **walnut** on its side still has a small abscessed node — which is the same thing that walnut meant many years ago.

You have to have earned your verbiage by time in the mud and the muck to be able to tell the difference between "pig," "hog," and "shoat"; to know that a **piggy female** is in late-term pregnancy; and to recognize that a **he-boar** is not a comic book character but a "pretty good ol' breedin' pig."

Thinking Hogs

Far more important than talking like a hog raiser is thinking like one. What causes the success or failure of any farm enterprise is not the cost of doing business, disease, or weather but the mind-set of the producer.

Even if you could be guaranteed that hogs would always sell for at least 60 cents a pound and corn would never cost more than $2 a bushel, you still would not succeed if your heart and mind were elsewhere. To be successful with hogs, you honestly have to like them.

Cattle are prettier, sheep are cuter, and there are few endearing qualities about an animal with a battering ram on one end and a manure spreader on the other. Still, the hardiness, vigor, and grit of the hog has to be admired, and its earning power is legendary. As I told my new bride, Phyllis, many years ago, any time hog prices go higher than 45 cents a pound, the smell of hogs is the smell of money. And of course, I must admit to a special fondness for the sight of a litter of baby pigs bedded down in bright, fresh straw.

Why Raise Hogs?

It might sound overly simple, but before you start buying, your first job is to determine why you wish to own hogs. When I was a young man, feeder-pig production was the best way to get into farming, and many people started out with a handful of acres, a small bunch of gilts, and some portable — largely home-made — equipment. Five gilts could become 50 in fairly short order, since they can be bred at 8 months of age to farrow at 1 year. Within eight weeks of farrowing, the producer has feeder pigs to sell. None of the other major livestock species can match this quick turn-around on investment.

Our first swine venture began with the purchase of a single crossbred sow with Hampshire markings we named Esmerelda. At the risk of giving you another clue to my age, I will add that we paid all of $35 for her. Our single-boar purebred Duroc operation has passed its 50th year, but we began it with the purchase of a single open gilt bought for $117.50 — and my share of the soybean check for that year was all of $120! As a matter of fact, most of the great livestock herds and flocks in the world can trace their origins to just one or two females of merit.

Unfortunately or not, small- and medium-sized farms can no longer stake their entire futures on just a single business venture or two. Hogs, with their ability to breed and farrow year-round, however, can fit into a great many different plans of farm diversification. Three to five sows farrowing twice a year can produce handy-sized groups of feeder pigs, herd replacements, or enough finishing hogs to fully utilize a 60-bushel **self-feeder**.

Identify Your Markets

Producers must be clear about the markets they wish to target with their hogs, the investment they can reasonably make in them, and their expectations for the outcome of their porcine venture. State fair winners have come from sow herds with five or fewer head, but those few were carefully selected and bred, were fed for optimum performance, and received their fair share of their producers' time and talents — producers who are true students of the art and science of swine raising. These men and women are breeders of good hogs, not suppliers of pork as a mere commodity. They honestly have feelings for the animals in their care, and though they raise hogs for family income, they are also doing things for the reason we, sadly, no longer hear about often enough: for the good of the breed. These producers know that what's good for the breed is good for their own hearts and souls, too.

Now, producers are more and more likely to be producing directly for end consumers and specialty processors. These are local folks — specialty-food processors, hog roasters, ham curers and the like — all buying for a product to sell or for home consumption. Our local farmers' market group, for example, now has a **Community Supported Agriculture (CSA)** wing that is using direct marketing to sell Berkshire-sired pork and whole and half carcasses to consumers in the St. Louis area. It is a project that should supply the buyers with a very special eating experience, generate a premium price for the animals being marketed, and give the participating farmers a year-round selling opportunity beyond the seasonal farmers' market.

I know of a small business that buys a few select hogs each month to turn into toppings for its line of high-end pizzas. There is a nearby producer currently marketing whole and half-hog carcasses to St. Louis restaurants, and another is now selling Italian and other specially seasoned sausage links at numerous farmers' markets.

A long time ago, noted farm writer Gene Logsdon told me that as long as there were public libraries I should never turn down writing assignments, regardless of subject matter. Similarly, with the backing of a good set of purebred hogs, I can think of no hog market closed to the independent producer who keeps an open mind and can employ the degree of

It's a warming sight to see a litter of baby pigs bedded down in fresh straw.

flexibility that has always been inherent in small farming.

A marketing opportunity is going to be where you find it now, and many are well away from the traditional Hog/Corn Belt. Opportunities will hinge not on huge numbers but on the producer's willingness and flexibility to meet those markets as they emerge. Many prospects will, no doubt, rise out of the growing interest in eating locally, out of respecting regional food traditions, and in presenting an alternative to the directions being taken by industrialized agriculture — something consumers increasingly fear.

2

BREEDS TRIED
AND TRUE

The three most popular purebred breeds of swine in the United States are the Duroc, the Hampshire, and the Yorkshire. They and their crosses are the backbone of the commercial pork industry. Their development and preservation was the loving work of generations of small and midsized family farmers.

Purebred livestock are vital for the integrity and sustainability of the family farm. Hybrid and composite breeding animals can become so complex in their structures that they are impossible to create on the family-farm level. Multiple crosses may be required to produce male and female lines, and they then must be mated in exacting order to maintain desired vigor. This is done in the name of creating improved herd output made possible by the increased growth, fertility, and disease resistance called **heterosis** or hybrid vigor.

The complex hybridizing schemes actually need a corporate structure to sustain them and lock the individual producers into off-farm sources for the all-important **seedstock** animals. It saps them of a great deal of power and control over their own fortunes, and most of the advantages of hybridized breeding stock actually fall to those who produce and market them. The purebred animal is the farmer's animal, one of the few herd-building basics (**inputs**) that still remain fully in the control of the independent producer.

Choosing Breeding Stock for Small Farms

Some years ago, I was part of a panel discussion at a farm show that addressed the task of selecting breeding stock for small farms with more traditional, outdoor swine-production facilities. At the time, they weren't exactly the forgotten sector in swine production, but the role they were to play was still being defined. The industry was still awash in volume producers, and the belief was that the seedstock sector would have an ongoing role in supplying breeding-stock needs.

That day's discussion is still quite relevant. The panel included a number of veteran breeders and writers in the field. Among their recommendations were:

- **Buy to match your breeding plan.** Avoid the unnatural extremes in type, as they are often quite short-lived fads. A number of breeders make it a practice to stick with hogs that are solidly middle-of-the-road varieties.

A young Duroc moving out and showing the desired long stride

- **Budget lots of time for the acquisition process.** You have to kiss a few frogs to find a prince, and you're going to have to look at a whole bunch of hogs to find the ones that will move your particular herd forward. Every generation should be as good as or better than the one that produced it. Small scale is no excuse to buy junk.

- **Look for old lines.** There is growing recognition of the value of old-line genetics, almost regardless of livestock species. It may be best seen in the quest for beef genetics bred for grass. The older swine-breeding herds and lines are the ones that retain classic breed character and have endured largely by inheriting a hardy constitution and a natural structure conducive to hardiness and durability.

- **Buy from deep, keeper-quality litters.** Choose from a litter wherein there are a number of siblings of very good quality.

- **Buy from herds that display a pig-to-pig level of genetic consistency.** Don't buy from litters and matings where the pigs display extreme differences in their type and quality. Be very wary of the ultraextreme individual that stands apart greatly from full and half siblings.

And there is something to be said for carefully selecting the producer from which you make your purchases. That was the last note touched upon in that panel, and one of the best-known breeders on the dais tied it up quite neatly. He said the producer to seek out is most likely ". . . over 50, breeds to middle-of-the-road type, selects carefully for reproductive performance, and never goes to a hog show!"

Modern Swine Breeds

A distinction of sorts is made between colored and white purebred breeds of swine. Although all of today's swine breeds are selectively bred for leanness, feed efficiency, and meatier carcasses and any full, purebred herds can do a quite creditable job producing basic meat hogs, the colored breeds are still considered to have the stronger "economic" traits. As a group, they are noted for their vigor, faster-yet-leaner growth, and meatier carcasses.

The white breeds, on the other hand, are stronger in the traits needed for successful pig raising. As a group, they produce milk and nurse and nurture their litters better than the colored breeds and tend to farrow pigs in larger numbers. They also have the docile nature you need to raise and wean large litters. On industrial farms, the focus has been to meld these traits into the hybrid überhogs, which produce pork with the consistency of ball bearings — really not a food item to be savored and enjoyed.

Confinement Hybrids

When the hog-raising sector began to divide so sharply between the great and small, there was also a division of sorts of its genetic resources. With a corporate mind-set that demands product consistency, the high-volume producers are now building on a scant handful of pure breeds and their crosses to create in-house hybrid lines rather like those seen in chicken-broiler production.

The pure breeds on which corporate farms have built their hybrid machines and that smaller operations continue to raise are the Hampshire, the Landrace, the Duroc, the Yorkshire, and its European counterpart, the Large English White.

A few manufactured, composite lines are seen now, such as White Durocs and some exotics with a dash to a heavy splash of the massively muscled, Pietrain breeding. Such efforts are designed to fold the traits of multiple breeds into a single animal. As I ponder these mixes, I recall one of my first lessons in swine breeding in the vo-ag classroom out at old R-1: "When you stabilize the traits in an animal to the point where they breed true for those traits, you have then created a new breed" were the repeated words of the instructor. In other words, if the newly created lines breed true, then they are truly new breeds and should be treated that way.

Many of the lines within the new breeds have been bred for generations for life inside a confinement building: Fat cover has been pared sharply, overemphasis has perhaps been placed on length and extreme muscle expression, and a most docile "confinement temperament" has been bred into them. These new breeds exist in the largest numbers of all of the swine breeds, are among the most familiar, and have even been used by some trying to establish more-naturalized production programs such as antibiotic-free rearing. There is some understandable concern for these modifications, however, as these animals do not fare well in outdoor production operations.

There are still lines within these breeds that can be considered "outside" hogs, but they must be carefully sought out and then subjected to rigid culling practices once taken up for those ends. Inside a confinement building or on a concrete floor, they are subjected to health, environmental, and genetic challenges very different from what a hog would encounter in its own world, with all four feet on the earth and both sunshine and rain falling on its back.

Popular Breeds: What's Old Is New

Joining the above group of confinement-picked swine breeds, there are now three other groups to be considered by veteran and prospective swine raisers.

The first group might be called the **heritage breeds**. Individuals in this group of animals have a long association with U.S. farms; among them are the Berkshire, the Spotted, the Black Poland, and the Chester White. Many of these breeds have fallen to near minor-breed status, and many have corrupted gene pools, as some unscrupulous sorts have sought to make them more extreme — heavier muscled, excessively lean — and similar in type to the confinement breeds.

A second group has come to be known as the **rare endangered breeds** and includes the Tamworth, the Hereford, the Gloucester Old Spot, the Wessex, and the Large English Black. A true, rare old-timer is the Mulefoot, which even in texts from early in the last century is referred to as "the old breed." These breeds all had roles of economic importance and were considered true farmers' hogs, although they were often popular only in certain regions of the country.

Then there is a third group that I refer to as the **porcine outsiders**. This wide-open group includes the Red Wattle, the Guinea Hog, the Choctaw Hog, the Ossabaw, the Russian or Wild Boar, and a couple that are somewhat swine miniatures, the Vietnamese Potbelly and the Kune Kune.

In simple and straightforward terms, all pigs are pork. The breeds were developed to function in different environments and under different sets of economic circumstances; over time their roles have evolved. I have even seen a growing number of hogs kept as companion animals. It is not a role that I believe is necessarily right for this species, although these animals are certainly quite sociable and often entertaining. Sadly, people are more familiar with the "racing pigs" seen at state fairs and in TV commercials than any of the other breeds.

We lost a number of swine breeds to extinction in the twentieth century, including the Ohio Improved Chester and the Sapphire. Others, such as the Berkshire and the Black Poland — the top two breeds for much of the first half of that century, fell sharply in numbers and importance. And the popularity of breeds is changing even now. Recent state fair shows for such breeds as the Berkshire and the Tamworth have been larger and more competitive than they have been in decades.

Testing has revealed that some breeds are more suitable for outside survival than others. They have the vigor and form that provides natural methods of protection. And certain breeds — including the Duroc, the Berkshire, the Tamworth, and the Chester White — are proven to produce more distinctive and flavorful meat than the pork that comes from the confinement industry. Others will possibly emerge to join this group. A small group of Red Wattle hogs was recently processed and submitted to a group of chefs as part of a Slow

SLOW FOOD IS GOOD FOOD

The Slow Food movement is a response to the fast-food world and its resulting decline in food quality and variety. The international organization states its vision as one of a world "in which all people can eat food that is good for them, good for the people who grow it, and good for the planet."

Originating in Europe, the movement has been a key player in preservation efforts on behalf of a number of livestock breeds and varieties. The group works to promote breeds known for superior table qualities, link independent farmers with chefs and the restaurant trade, educate consumers, and encourage an expanding role for local agriculture. Slow Food has a strong presence on college campuses and in large cities and can be accessed on the Internet at www.slowfood.com and www.slowfoodusa.org.

This Gloucester Old Spot gilt has good breed type, though she is showing some weakness in her topline.

Food group study. They tested favorably for both taste and cooking qualities.

Although the study is still in the preliminary stages, these breeds show great promise and should encourage universities and groups like Slow Food to carry out further testing for this and other breeds. Breed-identifiable meat marketing has proven to be a huge success with both Black Angus and Hereford beef, and this may be the future of the most delectable breeds of pork.

Colored Breeds

The Duroc is a truly American breed. It was originally known as the Duroc-Jersey, which arose from two types of hogs: the Jersey Red of New Jersey and the Duroc of New York. The Duroc ranges in color from dark, brick red to several shades lighter; it has smallish, drooping ears.

Because of its muscling, meat quality, and hardiness, the Duroc is often used as a terminal-cross male in a **crossbreeding** program. A **terminal** cross is one made to maximize growth and muscling when all pigs produced are to be harvested as meat animals. The Duroc still represents some of the most accessible of the genetics that make them ideal outdoor pigs as they are still bred in relatively large numbers. However, producers must carefully comparison shop for animals that represent the genetics with the old-fashioned grit and ruggedness of the breed itself. When crossed with Yorkshire, Hampshire, or Chester White females, the Duroc breed can create some top-rung F1 females for producing butcher stock and show pigs. See page 139 for explanation of the term "F1" and information about crossing animals.

HISTORY OF THE DUROC

A lively history of the Duroc can be found in *Modern Breeds of Livestock*, 3rd revised edition, by Hilton M. Briggs (1969). It seems that a man from Saratoga County, New York, named Isaac Frink obtained his first hogs in the 1820s from Harry Kelsey. While Mr. Frink was visiting Mr. Kelsey, he spied red pigs that took his fancy, so he bought some and took them home. The pigs had no breed name, so he called them Durocs, in honor of the famous Thoroughbred stallion owned by Kelsey.

The Jersey Red was well established in New Jersey by the mid-1850s. These hogs were extremely large, rugged, and prolific. But they lacked quality because they were long and rangy and had coarse haircoats and a large bone structure that detracted from their visual appeal.

It was the crossing of these two breeds that led to today's Duroc. Interestingly, however, the breed was built not on the East Coast, where it originated, but in the Midwest.

This Duroc gilt is showing good breed character and a nice level of muscling.

The Hampshire is of English origin, as are all breeds with *shire* in their name. It is a black hog with a distinctive white belt encircling the shoulder region. This color pattern dates back centuries, to a fad for breeding all manner of domestic animals for this striking color effect. Animals without a white belt encircling the whole body or white splashes in the upper part of the body are not issued a pedigree document. They are often termed "off belts" and are generally sold at discounted prices; however, they are not impaired physically by their lack of belting and still produce quality meat animals.

With erect ears and a trim appearance, the Hampshire is quite alert and vibrant in appearance. Some producers favor breeding and crossing hogs with erect ears, believing them to be easier to handle and drive. I believe you can tell more about a hog by observing its temperament than its ears, but I will admit that some lop ears once were so large as to present vision problems.

Hampshire boars are often used as terminal sires, and the Hampshire × Yorkshire gilt is the stuff of legend in the swine industry, known for its hardiness and productivity. Hampshires can impart rapid growth and exceptional leanness to their offspring.

Hampshire purebreds do not generally produce especially large litters, but the Hamp sow we once owned to produce F1 pigs when crossed with different Duroc boars maintained a weaning average of 11 pigs for eight litters. This is not a breed that has seen a lot of selective breeding for litter size in every herd, but many in the swine industry believe that the milk produced by a nursing Hampshire sow is nutritionally denser than that of a number of other breeds.

We live in an area that was once a real hotbed of Hampshire breeding activity and, as such, saw a lot of breeding trends within this breed. The Hampshire's role as a trend rider — the lead breed used to express the typically five-year-long trend in conformation and type — has not always been good for the breed. When seeking stock, beware of animals with

There is good bone under this gilt and a good topline.

HISTORY OF THE HAMPSHIRE

Hampshires are considered one of the oldest original breeds in the United States. According to the Hampshire Swine Registry, they probably originated from the "Old English Breed," a black hog with a white belt that was popular in Scotland. Hampshires came to America in the early 1800s from Hampshire County in England and were developed in Kentucky, where they were known for their hardiness, vigor, and foraging characteristics. They were also called McKay hogs because a man of that name supposedly brought them to America from England. In the early 1900s, the Hampshire breed "swept across the corn belt like prairie fire," the swine registry says. Hampshires were also used to produce the famous Smithfield ham.

excessively heavy muscling and too little fat cover. These animals can often be too extreme and unnatural in form, and among stockpeople there are questions of genetic purity regarding some of the more extreme animals currently being touted as a part of the breed.

The Spotted, once known as the Spotted Poland China, is the colored breed with the black and white spotting pattern that gives it its name. This breed also has drooping ears and is known for its good mothering abilities. The carcass is long and trim and until recently was not as heavily muscled as the Hampshire or Duroc.

The Spotted breed was perhaps more popular east of the Mississippi River than in the West, and its population is now sharply down in numbers even in such former hotbeds as Indiana and Ohio. Spotted sows bore some of the largest litters of the colored breeds, building a reputation for producing classic feeding hogs and cornfield hogs.

Spotteds share a common ancestry with the British heritage breed, the Gloucester Old Spot. In the United States, both breeds have been tracked by the same breed association. In England, where it is also known as the orchard hog, the Spotted has enjoyed the support of the royal family. There, aficionados claim that it has been bred outside confinement for more than a hundred generations.

The Spotted is a breed that has changed dramatically in appearance over time. The Spots of my youth showed more background white coloring; had big, roomy ribs with lots of body capacity; and were really good mama sows. I fear some outside blood has entered this breed, and you should select carefully for type consistent with the breed's traditional character and performance role.

I would like to see it returned to its more traditional and durable type — that of a well-performing cornfield hog. That type still exists and should be sought out and used more in modern breeding programs.

THE SPOTTED BREED

The Spotted breed offers a different choice for those who need both the hardiness of a black-colored breed and good litter size in their black/white/red, three-breed rotational crossbreeding plans. An example of such a cross would be a Spot/Duroc/York.

The Berkshire is an old-line breed now enjoying a resurgence in popularity. In the 1930s, it and the similarly colored Black Poland China led the nation in number of recorded pedigrees. It is a largely black breed with white points and splashes found predominantly in the face and lower half of the body. It has rather short, erect ears.

At one time, the breed had a very short, upturned snout, but selective breeding in recent years has removed this trait, which many thought was conducive to feed waste and respiratory problems. The Berk is a well-muscled breed noted for both leanness and high-yielding carcasses. Pork from pure- and high-percentage-bred Berkshires is noted for exceptional table qualities, including quite large loineye areas and finishing qualities that give the pork something akin to the marbling in high-grade beef. This pork is especially valued in Japan, where it sells for a premium.

As with Duroc and Hampshire boars, the Berkshire is often thought of more as a terminal sire. The females make decent mothers, however, and Berkshires are hardy and durable.

The Black Poland China (known as simply the Black Poland on the farm) is nearly identical to the Berkshire in its color pattern but has drooping ears. This hog has perhaps a flatter and longer topline than the Berkshire does and produces exceptional meat qualities. The carcasses are lean and well muscled and yield a good dressing percentage that includes exceptional hams.

The Black Poland also is widely known as one of the most durable of all of the swine

This bred gilt is a bit behind in her type, and I would like to see her with more bone and more length. She is showing the expected udder development for a pregnant gilt.

breeds. At one time, this breed's association herdbook could lay claim to more females at eighth parity (in the eighth litter) and beyond than any of the other major swine breeds. Black Poland females are often underrated as brood sows, but they can add real punch to a swine farm where hogs are kept outside on pasture or in a drylot and a very hardy black breed is needed to work into a crossbreeding rotation.

The Black Poland and Berkshire are both gaining in popularity as sires of show and youth-project pigs. Though they have slipped sharply in numbers in recent years, they nevertheless have a bright future as breeds that have much to contribute to the production of lean, well-muscled market hogs.

White Breeds

The white breeds remain useful in both their pure state and in various crossbreeding rotations, although some white breeding lines have been badly co-opted by overbreeding for life in confinement or to impose upon them a more extreme meat type. I would like to see them selected more for their vigor and durability rather than just for traits that turn them into pig machines with a fast turnover.

I remember reading accounts of a 100 percent pure Yorkshire commercial swine herd based in Colorado a number of years ago. This practice is nearly nonexistent today, and it was rare even then to have a totally purebred herd producing butcher stock. The herd remained pure, in part, because it was in an out-of-the-way location, making it hard to access varied sources of replacement females. The producer selected with an eye for balanced meat type, all female replacements were home raised, and outside purchases and buying trips were limited to a new boar or two every couple of years.

The Colorado operation sets a good example for the new generation of raisers. It proves that breeders can select for reasonable meat type, growth, and reproductive performance within an individual breed without sacrificing any of the breed's true strengths. It took us many years of selective breeding with our Durocs, for example, but we did advance the vigor and mothering abilities of our sows to the point where we were consistently weaning 8 to 10 pigs per litter.

Remember, the goal is to achieve optimum performance from a given set of hogs and facilities. Maximum levels of performance in any type of production agriculture are seldom, if ever, truly cost-effective, as they must be supported by fancy housing and more costly rations.

When selecting for white breeds, you should evaluate carefully the genetics available now to be sure to acquire animals that are absolutely pure. Make sure the history of the line boasts something other than life in a confinement facility. To this end, eyeball them carefully for structural soundness, true vigor, moderate muscling, and internal dimension. You need a big barrel of a body on these hogs and should perhaps seek out individuals with older, deeper pedigrees. There were still some good genetic pieces in place as recently as the early to mid-'90s. The genetic-selection methods are still available, and they badly need to be taken up again. The task is to find that just-right gilt with which to begin.

A lot of the current genetics are sold out to confinement production. The range producer must study pedigrees and herd histories to find lines that will work outside. General methods for eyeballing and evaluating breeds are described beginning on page 59.

The Yorkshire lays claim to the title of the mother breed in its advertising, and in most years it leads the nation in total registrations. Large litter size at both birth and weaning is a breed standard and evidence of its superior mothering and milking abilities. Many people recommend this breed for a youngster considering a gilt-and-litter 4-H or FFA project. A Yorkshire gilt will provide the youth with an opportunity to work with a great many pigs and will be easiest to find; Yorkshires are among the most widely available of all of the purebred swine breeds.

The Yorkshire is all white, with sharply erect ears and dark eyes. It shares its origins with the English Large White — both are recorded with the same breed association in this country. The Large White is perhaps the definitive confinement breed.

Yorkshires also figure prominently in a great many of the more popular crosses, including some with the other white breeds, to supply highly productive female replacements for commercial operations. The Duroc × Yorkshire is growing in popularity in the Midwest as a way of creating an even more durable carrier for white genetics. The smallholder with simpler facilities will be served best by what are termed "American" Yorkshire breeding lines, which are more selectively bred for life in outside lots and pens.

The Chester White is an American-developed white breed that originated in Chester County, Pennsylvania, with the name Chester County White. It has a medium-sized frame and drooping ears. It is probably an underutilized white breed, considering its hardiness, which makes it appropriate for producers working outdoors with simple facilities.

The sows are good mothers and become pregnant very quickly following weaning, a very important economic consideration. Not only does this reduce operating costs, but there is also strong statistical evidence that the more quickly a sow breeds back following weaning, the larger her following litter will be. We have owned Chester sows that farrowed three litters in one 12-month period, and we have owned many that maintained lifetime litter averages of 10 or more weaned.

Chester White pigs are a bit smaller at birth than some others but are generally quite vigorous. In large litters, the pigs are more likely to be smaller. Purebred Chesters are

YORKSHIRE HISTORY

The Yorkshire originated in England; the first of these animals to cross the sea was brought to Ohio around 1830. The American Yorkshire Club says that the breed initially failed with farmers because the hogs were slow growing and had short, pugged noses. A man in Indiana is credited with showing farmers the advantages of Yorkshire breeding stock, which include larger litters, great mothering ability, and more length. Once farmers realized these benefits, the breed gained in popularity. The first Yorkshire registered in the United States was Clover Crest, a boar imported from Canada.

Today, Yorkshires can grow at a rate of more than 3 pounds per day, with an exceptional feed-conversion rate. The goal of the Yorkshire breed, says the Yorkshire club, is to be the source of durable mother lines, contributing to longevity and carcass merit.

also a bit slower growing than some other purebreds — trailing the norm by three to eight days to harvest weight — but still achieve 230 pounds in a respectable number of days of age. University testing verifies that one of their real strengths lies in what they can contribute in crossbreeding. They carry the typical strengths of the white breeds — milking, mothering, and litter size — while allowing the growth and carcass traits of other breeds in the cross to shine through.

Yorkshire

Chester White

The Landrace is considered the longest of all the more popular swine breeds and is also known for its distinctive long, drooping ears. Selective breeding has greatly reduced Landrace ear size of late, however. This white breed also produces very large litters of pigs.

A bit finer boned than other breeds, the Landrace is more often used for indoor confinement breeding. Among the most docile of swine breeds, it is frequently used in cross-breeding to add to both litter size and carcass length of the resulting offspring. Separate Landrace strains are bred throughout Europe and Scandinavia, where its temperament is valued in the close-confinement housing commonly found there.

If penned outside, some Yorkshire and Landrace sows may need a bit of extra feed to maintain body condition. They also seem to perform best when kept in groups of animals with similar breeding.

Rare and Minor Breeds

Interest in the minor breeds has grown remarkably in recent years. The reasons for this are twofold. Heritage breeds are old breeds valued by consumers and raisers for the same reasons that people value heirloom poultry or produce. Heirloom tomatoes and heritage hogs are reminiscent and representative of days of yore; they are often tasty and suited for small farms. There is also a growing regard and demand for local and regional food items and food supplies, a market for which the minor breeds were developed. For example, I'm from the Show-Me State, and when we want really good pork, we feel it must come from a hog with a lot of red on its hide and in its genes.

Heightened consumer awareness and concerns have greatly enhanced the prospects for raisers of the Tamworth and Berkshire breeds. These breeds have quite a reputation for

Landrace

TRAITS OF POPULAR BREEDS

Livestock judges are trained to never speak in negatives. One animal may be better than another, but all hold to a certain norm that has kept them in place. They have different roles to play and thus present with different strengths. Below are the traits that make each breed unique.

Breed	Traits
Berkshire	Premium table-quality pork
Black Poland	Especially durable; good for crossbreeding in an outdoor swine operation
Chester White	Hardy white breed; breeds back quickly
Duroc	Top eating quality; hardy enough to raise outside
Hampshire	Offspring grow rapidly, with exceptional leanness
Landrace	Large litters; docile; usually raised indoors
Spotted	Excellent mothering ability; large litters
Yorkshire	The mothering breed; large litters; good for indoor confinement

producing pork with distinctive cooking and eating qualities. In fact, the Midwestern state fair breeding-stock shows that feature these animals are now among the largest and most competitive in the country.

As the Tamworth and Berkshire breeds have gathered interest, they also have bene-fited from renewed efforts at selective breeding for improved meat type and performance. They had minor-breed status, and many thought that all they had going for them was a bit of history, but that history of hundreds of years of serving American farmers and their patrons is now very marketable, and these breeds are being produced in an infrastructure quite separate from that of the rest of the swine industry. They often garner substantial price premiums through marketing programs that promote animals that are at least 50 percent Tamworth or Berkshire.

There is a widely known buyer in the eastern Corn Belt now paying a premium for Tamworth gilts to be bred with his Russian boars and the shoats from such crosses. Some of these animals are then used to stock hunting preserves, and others are marketed as pork in the fast-growing game-and-exotic-meats trade.

In a few instances, even the purebred Russian boars (also called wild boars) are now being farmed. They are most hardy animals, raised on range year-round, and while litter size is quite small, they produce a meat animal that is in high demand. Housing needs are modest, although fencing must be stout and should be reinforced with a good electric fencing system. Research your market outlets thoroughly before beginning with these animals, as they have only a modest demand and one that can be quite seasonal in nature. They are one-litter-per-year animals, their handling should be kept to a minimum to reduce shock and stress, and if at all possible, they should be processed locally, as they are hard to handle and ship.

Heirloom (Endangered) Swine

Another group of swine breeds found in the United States are the heritage breeds — hogs that were more common in an earlier day and have continued into the present in small and often widely scattered populations. Not many years ago, the **Mulefoot** breed, for example, was believed to exist in just three small herds.

Heirloom breeds are still valued for a number of reasons, including the simple fact that they represent some of the hardiest of all the swine genetics. As a group, these animals are quite naturally lean, adapt readily to a wide range of environments, and are among the best choices for a pasture- or range-based production system.

The Tamworth is a light red breed of English origin with erect ears. This is probably the most vigorous of the swine breeds and one that has an unearned reputation for having a bad temper. For instance, one of the breed association presidents who lived in our county regularly worked in pens with Tamworth sows and very young pigs and always exited unscathed, even though he was encumbered by a debilitating disease.

With a frame that is slightly smaller than some other breeds, the Tamworth has remained hardy and retains many of the characteristics of what used to be called the "range hog." An acquaintance of mine tells of once unloading 10 Tamworth sows and 101 pigs into

THE MULEFOOT

A very minor and truly endangered swine breed is the Mulefoot. Today it exists only in the black-color phase, has drooping ears, and has a solid hoof like a mule — not the cloven hoof seen in other swine breeds.

One of the last herds of the Mulefoot breed endured because it was isolated on islands in the Mississippi River near Louisiana, Missouri. The herd lived primarily on its own, foraging in the summer until rounded up in the fall for slaughter by the late R. M. Holliday, who fed grain in the winter and selected carefully. Mr. Holliday's herd was dispersed after his death. Late in life, he got to see a renewed interest in his breed of choice. Its survival is a testament to the hardiness of the breed.

The Mulefoot is a bit fine boned and lacks modern meat qualities such as well-muscled, high-yield, long carcasses. Although critically endangered, its cause has been taken up by The Livestock Conservancy and some affiliated breeders. This should help ensure that the Mulefoot will provide truly alternative genetics for the swine industry.

Legend has it that the Mulefoot breed was resistant to the once-dreaded swine disease hog cholera because it had no cloven hoof, which prevented the disease organism from entering the hog's body. This was a good story, but it had absolutely no basis in fact.

The Mulefoot is a very hardy breed, perhaps closer to the wild hog than any other current swine breed. The red and spotted color phases of this old, old breed are now considered extinct. One text from the 1800s refers to it as an old breed of swine even back then. When crossed with other hogs, the mulefooted trait might appear only on the front or back two feet or not at all.

a large wooded pasture; he left them with self-feeders, but they were otherwise largely unattended for several weeks. When he returned to gather them up at weaning age, the sows had safely raised all 101 pigs. The Tamworth sow lies down by first dropping to her front knees and then shuffling down in a way that keeps her from crushing her pigs. In fact, Tamworth sows are deemed to be the best mothers of all the colored breeds of swine.

The Tamworth was once used as an example of what used to be called the "bacon breed" of swine. It is a very lean hog, and some of its lines are too thin. Unfortunately, like many of the other minor breeds, its type is a bit stuck in time genetically. Tamworths need more producers to take them up and selectively breed them — to maintain the traditional strengths of the breed but also to improve growth and muscling, which translate into better carcass yield and more cuts that are prime hams and loin.

The Wessex is a rarely seen breed with the conformation of the Landrace, including the drooping ears, and the distinctive black and white color pattern of the Hampshire. Many believe it to be hardier than the Landrace, though it tends to produce somewhat smaller litters. It is a valuable breed lacking in breeders and numbers. There may be more of these animals in Canada, although the Wessex is valued by many breeders in the United States who raise hogs outside on pasture or in drylots.

The Tamworth is a good breed for the pasture or range producer and has an unearned reputation for having an edge to its temper.

Wessex

Although once important in the United States, the Wessex are not included in The Livestock Conservancy endangered list, as they are in such few numbers as to be considered extinct. The last I saw offered for sale were in Canada, but like the Mulefoot, I believe they're still here and may have a slim chance of survival.

The Large Black is another minor breed with a stronger Canadian presence but sometimes heard about in the United States. A prolific breeder and good pig raiser, this breed is of English origin. It is all black and has large, drooping ears. Although it isn't as long or trim as other breeds, the Large Black represents some enduring swine genetics that could be brought back into use. Like other minor breeds, the Large Black has been neglected by the industry, supported only by a few breeders, and subjected to higher levels of close

breeding; it endures, nevertheless, due to its innate vigor.

The Hereford is an especially attractive breed, with a red and white pattern mimicking the distinctive color pattern of the Hereford breed of cattle and drooping ears. Its frame is a bit smaller than that of some other swine breeds.

I have seen purebred Hereford hogs place well in recent market-hog shows. At one, a purebred Hereford gilt topped the always competitive class of 240- to 245-pound market hogs — the prime marketing weight for these hogs. Friend Ron Macher, publisher of *Small Farm Today* magazine, used to buy purebred Herefords to feed out and process into extra-lean pork sausage, which he marketed directly as a gourmet product. This is a pretty breed of hogs promoted for practical ends.

Large Black

Hereford

Old-Style or Southern Hogs

Russian or wild boars are the bona fide, naturalized variety; they are real-deal pork — straight from nature. Today's feral hogs sprang from domestic stock that have somewhat reverted to wild type and have existed in the United States since colonial times. With a moderate climate and access to an array of potential food sources, a hog is a self-sustaining animal.

My grandfather was raised in northern Alabama in the early 1900s, and the most feared animal found in the woodlands there was the range sow with nursing pigs to protect. It was common practice to free-range hogs then, and once freed, they lived a catch-as-catch-can existence. In the late spring, each owner tried to catch his pigs in a pen trap baited with a bit of grain and then give each pig the owner's identifying ear notch.

In the 1990s, when the price paid to farmers for live hogs plunged to less than 10 cents a pound, many thousands were simply turned loose into the woodlands surrounding farmlands throughout much of the country. One of the reasons for the release was to create populations of a "new" game species for hunting; however, hogs are not a part of most natural populations. Feral hogs grow rapidly in number, can harbor diseases risky to other species both wild and domestic, can overtax natural systems, can damage growing crops and pastures, and can harm ponds and other

Southern hog

Red Wattle

water systems. The "pigzilla" that was featured in so many newspapers and countless e-mail attachments a few years ago actually appeared to be a Duroc boar that was fairly typical in size for a domestic hog allowed to grow for several years. Hog breeds claiming some ties to the wild include the Red Wattle, the Guinea Hog, the Choctaw, and the Ossabaw or Ossabaw Island hog.

The Red Wattle breed suffered a bit of a scandal some years ago when the wattle trait was bred onto hogs with a far different background. You must be very conscious of selecting for true breed type and character when buying breeding hogs now. Sometimes even pedigree papers are no assurance of genetic purity. There is an old adage in the livestock sector that holds that a pedigree is only as good

as the man or woman who hands it to you. For example, Chester Whites that look like Landrace crosses or Tamworths with Duroc type or color should raise all sorts of red flags.

The Choctaw is believed to be descended from Spanish stock brought to the South by early Spanish explorers. It takes its name from the Choctaw people and was carried by them and other tribes through the South and Southwest. Some are black hogs with a few white markings, some have erect or slightly tipped ears, and a few may show wattles. They are fine boned, have a quite light mature weight, and may even show the "hooves" of the Mulefoot. You might find similar-looking red- or sandy-colored hogs described as Choctaws because the standard for this breed has not yet been set.

The Guinea Hog has an unclear origin, although many believe that it can be traced back to West Africa and the time of the slave trade. They can vary greatly in size, are vigorous by nature, have a truly coarse haircoat, are black in color, and have erect ears. They exist in very small numbers but claim a great deal of genetic variability for such a small population, which means they are not overly inbred.

The Ossabaw takes its name from Ossabaw Island, located off the coast of Georgia. They are descendants of hogs left there by Spanish explorers four centuries ago, a self-perpetuating meat larder to be tapped later. The original population is still confined to that island as a health measure, but some two hundred animals were brought to the mainland in the 1970s. They were also imported to Missouri for medical research in the '90s and can be found on several breed-preservation and period farms across the Midwest. These are the basis for the breed's presence in the continental United States now.

These hogs grow larger than their still-feral island cousins that mature at around a hundred pounds. They have a very bristly haircoat and are found in black, red, and sandy-tinged colors, with some instances of white spotting possible. They are beginning to carve out a bit of a marketing niche due to the quality of the pork they produce, but like the other minor breeds described above, they are riding along on their novel nature and relative smaller size.

As you read the descriptions of these breeds, something may have begun clicking in the back of your mind: They share the physical description and lore of what your grandparents and mine called "razorbacks" or "ridge runners."

Guinea Hog

Ossabaw

Many years ago, I stood with Dad and an elderly gentleman and watched as a drove of "Southern" hogs was unloaded and penned at our local sale barn. Just above the pen where they were driven hung a sign that would be worth a small fortune now as a bit of American folk art. It read, No Ruff Sothern Hogs.

One of the largest animals in the group reared up on a gate, tore down the sign, and began to chew it to pieces.

"You don't suppose that hog could read, do you?" the old gentleman asked.

"That, or he was raised around here close and took offense at the name calling," Dad replied.

As I write this new edition, much is being made of "Southern" hogs, and there is a movement to establish or at least legitimize a razorback breed. The swine-breeding sector worked for many long years to shake any trace of that image, and many in that group see the razorback as little more than the porcine version of the cur dog.

Rolled up in all of the attention to these animals is, I believe, a number of matters both great and small — the growing interest in smaller versions of established livestock varieties, the movement to preserve heritage livestock breeds, the quest for easier-to-maintain animals, a desire for animals that produce a more distinctive pork product, the growing interest in regional and local foods, and the long-established fascination with the **exotic** and unusual.

There is nothing wrong with any of these expressed points of interest, but you should realize that these animals come with some very big ifs and potential limits. That a breed may be a passing fad is certainly a concern when buying breedstock. Only the Red Wattle breed seems to have a breeder group behind it doing any sort of thinking beyond the immediate pricey and dicey "breeders' markets" that raise and promote these breeds. Such markets are driven by short-term interest in the current hot thing, and although any kind of interest is encouraging for hog raisers, the end purpose of these animals has yet to be clarified.

These animals have existed outside any breeding structure, not just for generations but possibly for centuries. In their breed descriptions, we see striking variability within the breeds and equally striking similarities among the breeds. Some preservationist groups, such as The Livestock Conservancy, have brought needed focus on these breeds, but where is the long-term thinking about them? There needs to be a real meeting of the minds as to where these breeds should be genotypically and phenotypically.

Do we need a razorback breed? No. Emphatically no. But we do need to preserve the breeds and lines that are here. We do need to respect their roots and the roles for which they were created. These hogs have yet to find their long-term niches. It may be as roasters or as the staple behind an element of regional cuisine. It may be as a meat animal for the growing number of microfarms or as part of the set of genetics needed for modern range production. Only time will tell.

Minipigs

A couple of breeds that are growing in number but that do not fit any practical niche are the Vietnamese Potbelly and the Kune Kune pig. Their greatest appeal for the moment is as "miniature" swine, but they were developed as meat animals for very constrained environments. They were raised by families who had limited feedstuff for the livestock, who had to eat all meat when it was fresh, and for whom every bit of food counted. These and other environmental restrictions dictated the breeds' sizes.

The Vietnamese Potbelly has come to epitomize all that is good and bad with the exotic animal trade. Early on after their appearance in the Midwest, the animals traded for very modest prices; but as interest grew, they took off like the Hula-Hoop, and prices went through the stratosphere. Not many months later, they were being given away.

The little pigs are black with white points and erect ears, but they have been ruthlessly exploited. Unscrupulous breeders mated them with smaller lines of other swine breeds, and soon there were white and red varieties and varieties that grew much larger than the previously accepted size for potbellied pigs. At the peak of their demand, the rarer white and red pigs would sell for five figures in the Midwest, but as their numbers grew their prices plummeted. Most folks in rural Missouri now have stories of the pigs being given away or passing through local sale barns for just a few dollars per head. A great many here now, too, know what one of the pigs tastes like.

EXOTIC BREEDS: WHAT FOR?

The exotic breed sector has a hard question to answer and one that must be answered soon: What is the ultimate use of these animals?

They were initially developed as meat animals and will soon saturate the "breeding" market that exists for them. If preservation work succeeds, the "rare" or "sellers'" market it fosters will go away. They're too small to ride or pull a plow and need only a few breeders if they are merely to be maintained as a genetic presence.

What is the long-range vision for them? Where are the guidelines for breeding them for better and more production? What market niches will they fill 5, 10, or 20 years from now? I try not to be too cynical about some of these swine "exotics," but as I said, I'm from Missouri, and they still have much to show me. Are they going to be developed and bred up to be an alternative meat animal?

The modest-sized whole and half carcasses of exotic breeds might make them a good choice for today's smaller families. They may produce some exceptionally flavorsome pork or lend themselves to certain cuisines. What do they taste like?

I watched the potbellied pig fiasco up close in the 1980s and '90s, and it was filled with crooks, high winders, and hot-tea drinkers that knew little or nothing about mud and manure. A lot of good people lost a lot of hard-earned money with them, and some never came back from that loss.

I don't like myself when I become skeptical, but until I see a Choctaw or an Ossabaw hog win a market-hog show in the Midwest, I am going to continue my pessimistic ways. Eventually, they will have to play by the same rules as the Hampshires and Durocs and their producers. To do this, they must be moved along in that direction now. This means **selective breeding** for litter size, economic traits, and a consistent standard for breed character.

All pure genetics — genetics that will breed onward in a reasonably true and predictable way — have value. It's true that these exotic breeds may represent the only fallback position in the event of a genetic disaster; however, they are not show horses and gaited ponies. Sadly, right now they are valued mostly as hobbies for the wealthy.

These hogs need some showdowns in the show-ring, true-type conferences, a pool of data proving their worthiness and the specific roles they can play, and producers that see them as livestock with a future and not just two hundred specimens with a past.

POTBELLY IN THE POT?

At the peak of their popularity in the region, I was sitting in the local vet's office (he was a high school classmate of mine), when a lady brought in a number of the little pigs for some basic health treatments. She engaged the vet in a discussion about the Potbellies and their future beyond the then red-hot breeders' market.

He replied that they were hogs and that they could be eaten; in fact, he had eaten them when in the service in Vietnam. A bit hesitant and fearful, she still pressed him for details. They would flush one from the brush, he told her, shoot it, field dress it, and sit down and wait for 10 to 15 minutes. Now really fearful, she asked about the waiting time. He replied that it generally took that much time for the next company with a flamethrower to pass down the road.

Vietnamese Potbelly

The collapse of the Potbelly came about for a number of reasons. With a steady and richer diet, they naturally grew larger, but because the breed had been tinkered with and crossed with other breeds, after a year or two many found themselves with porcine miniatures in the 200- to 300-pound weight range. And as is the norm, numbers grew greatly as interest ebbed and demand fell. The resultant drop in interest put thousands of these pigs into pig rescues and animal shelters across the country, and many more were euthanized.

Interest in and demand for these little guys continues in some parts of the country. They are now being propagated by a corps committed to safeguarding breed integrity and breeding for the desired smaller size. These breeders must take care, though, that they don't take the desire for smaller size too far; it could conceivably compromise the animal's ability to thrive and reproduce.

The Kune Kune is a small spotted pig, and the varieties I have seen are black and white spotted or sandy red with black spotting. They have erect ears and short, pug-nosed faces that remind me a great deal of the Berkshires of many years ago.

Kune Kune

Smaller-sized pigs such as the Potbelly and Kune Kune are more easily transported, a trait that was especially crucial in locales such as Polynesia, where the supplies and livestock needs for a new settlement had to be carried by canoe and outrigger. These pigs can live in harsher landscapes, where their food supply is limited and their size is economically justified. These miniature swine were and still are hogs in every sense. Unfortunately, they are being raised in this country primarily as gimmicks to sell to others.

Why Purebred Hogs Instead of Hybrids?

There are not as many purebred hogs today as there were even 20 years ago. I recently came upon a farm magazine published in 1970, two years after I graduated from high school. In it was page after page of advertising for purebred breeders' businesses that slipped away just before the turn of this century. In that year, there were at least 15 purebred spring and fall auctions within 25 miles of our home.

Those purebred animals sometimes went to other purebred producers but more often to local family farmers, who used them to formulate crosses to produce feeder pigs and butcher stock. The producers could assemble animals of desired breed, type, and conformation to create the hogs that would perform the best on their farms or for their customers. They crossed the purebreds to gain the fullest possible advantage of heterosis (hybrid vigor) in the resulting offspring. Most hybrids are stronger than their purebred parents; the result of two distinct bloodlines coming together is a stronger, more durable animal. But the traits that are expressed are also less predictable because they are two different breeds.

These days, hybrid vigor is accomplished almost entirely by seedstock companies that sell crossbred and composite breeding animals. Hybrid hogs are really nothing new, and private and U.S. Department of Agriculture (USDA) research as early as the 1930s produced the various Minnesota, Montana, and Lucie hybrids of that era, which exist to this day.

Many of the modern so-called hybrids are quite complex in their structure, and because the strains are so closely managed, they lock producers into joining a contract program using a whole phalanx of company hogs to maintain even a semblance of hybrid vigor. It is an expensive and complex practice that requires much genetic fine-tuning and extensive housing and produces some animals that are not as productive as purebreds. It's a system that could shut individual producers out of swine seedstock production in the same way that independent poultry producers were blocked from producing the now-favored broiler and laying-hen strains.

PUREBRED SEEDSTOCK AND THE INDIVIDUAL PRODUCER

Without the purebred hog, the genetic resource for the people, swine production would lose the capacity to sustain itself. Selective breeding can shape purebred lines to fit local environments, changing economies, and the varying nature of consumer demand. We have gone far too long with agribusiness trying to dictate to farmers and smaller producers which animals they should produce and want.

Choose a Breed Type to Match Your Farm

Breed choice can be based on many things and will vary from individual to individual. Our long association with Duroc hogs, for example, came in part because my grandmother and my wife, Phyllis, both had a deep fondness for the classically colored red hogs.

The Duroc breed crossed complementarily with other breeds available in our immediate area, had a long history as a farmer's hog — a hog that is practical and easy to keep — and we could easily access sources nearby for foundation and replacement animals. Long ago I was cautioned that if you bring something in from 1,000 miles away, you may have to haul it back every one of those 1,000 miles to get it sold again.

The iconic hog of the Midwest is the sandy red pig with a sprinkling of black spots, the classic "cornfield hog." The colored hide bespoke hybrid vigor and genetic input from two breeds known for winter hardiness and practicality, the Duroc and the Spot. The great task for this new generation of producers is to restore the traditional roles of various pure breeds and revert their carcass and muscle type back to the shape for which they were originally bred.

Nothing in the way of true productivity (mothering, litter size, milk production) will be lost by breeding a Landrace to the standards once held for this grand breed — one of the white mother breeds. The Hampshire is a growth and muscle breed but one that should not be bred to the absolute extremes seen in

WILBUR LIVES!

It is very much a part of human nature to want to make pets out of all sorts of creatures, and few others are as personable and engaging as the young pig. I know of many instances where hogs have been kept as companion animals, although they were not quite like Siamese cats or Yorkie lap dogs. Pet pigs must be handled differently — a "lap pig" will hurt you!

Many years ago we had a shoat pig that suffered bouts of gastric ill, and he just would not grow. He made his home with us for a couple of years, and Dad even fixed him a sleeping area in the machine shed. The little guy supervised any number of shop projects, and Dad took great pleasure in walking by with him at heel as I was extolling the virtues of our breeding program to potential boar buyers. An old-timer finally showed me how to cure his ills with draughts of buttermilk and charcoal, and I actually got him to start growing again.

The truth is no one really knows how large a domestic hog may grow. University studies have shown that hogs continue to grow after 10 years, and most such experiments are then discontinued due to the lack of space with all the subjects still growing! At the few state fairs where "big boar" contests continue, boars in the 900- to 1,000-pound weight range are not uncommon.

A couple of years back, a woman stopped at our local farmers' market seeking a pig to be a pal to another hog that was a companion animal to the horses in her stable. Dad gentled a number of our breeding hogs to the point where kindly rubbing their bellies with a stick or the toe of a boot would cause them to flop at your feet like a beagle puppy — a puppy weighing several hundred pounds!

animals produced for corporate farm confinement and the show-ring. Breeding should not compromise the animal's ability to function naturally and in its traditional role as a versatile animal. Any cross should be one with practical merit, and the Hampshire × Landrace is a classic maternal cross.

I can remember a generation of breeders — real hog folks — who could recite pedigrees of popular hogs from decades earlier, who knew breed strengths and weaknesses and how to work through and with them, and who long ago realized that the hog has a role to play in the field, on the plate, and in the processing line. They knew the good hogs, how they were put together, who had them, and who screwed them up. We need — desperately need — more of those kinds of producers.

Pigs as Pets

If your heart is set on having a pet pig, weigh the decision carefully and take plenty of time for the selection process. A lot of junk Potbellied pigs have passed down the pike in the last few years. I have a friend who is an animal control officer; he's been asked to put down little pigs that weighed well in excess of 225 pounds. A young pig is puppy-dog cute, but even one of the smaller breeds can grow substantially, has unique food needs, and may not get proper care; not every suburban veterinarian is going to be up to speed on swine health.

I try not to take a negative view about pigs as pets, as I realize that many people are willing and able to invest the time and resources it takes to maintain one of the swine varieties as a companion animal. If you decide to look into it, thoroughly investigate the breeding line of the porcine miniature you are considering by doing the following:

- **Ask pointed questions about genetics.** What can the breeder tell you about purity and the size of all the seedstock in earlier generations? Are any of the related pigs on-site? Can they be seen? Has size been kept small through excessive inbreeding or limited feeding?
- **Ask about temperament and tractability.** How easy is it to work and live with animals of that line?
- **Ask about the seller's rates and policies.** Is there a return policy subject to a health inspection by a veterinarian?

I chose the farming life largely because of my love for animals and birds and a desire to create a life and career with them. In my nearly 60 years at this calling, I have formed great attachments to many of God's creatures and have a great many pleasant and bittersweet memories.

The first Duroc female we owned came to be called the "Granny" sow for all of her offspring on the farm. After producing 10 litters, the years began to show, and there were better daughters and granddaughters waiting to fill her place in the herd. When we put her on the truck to town, it was a bittersweet moment, but she had had a good life with us. She had fulfilled the role for which she had been bred, and her selling price would be one last contribution to sustaining her offspring and our family. A half century later, I can still remember the night I sat close to Dad's side and bought her. The lessons we learned with her are the ones that have helped to make this very book possible.

This hog is being walked around the farmstead to become better conditioned for the show-ring.

3

RAISING HOGS FOR THE FAMILY TABLE

Most folks start thinking about hog ownership as a way to provide quality meat for the family table at moderate cost.

When you raise your own hogs, you select them, feed them out to an exact slaughter weight, and direct the processing. Feeding out a hog for slaughter is finishing a hog, which gives you an assurance of quality and wholesomeness that you can have in no other way. And along with quality control, there is much that you as a home finisher can do to contain costs. Granted, a pig is not all chops, but when it is raised and processed to order, you can expect to maximize the cuts and quality you and your family prefer in meat and meat products.

The Facilities

Where do you begin? Chances are you'll start with one or two feeder pigs. The first thing you'll need to think about is where to raise them.

Pig raising is simplest and most efficient in climates without great extremes in temperature or precipitation. The **growout period** will normally be between 90 and 120 days,

depending on the starting weight of the pig, and in many places fits into the spring or fall season, to avoid the weather extremes of a Missouri summer or a Maine winter.

The finishing pen — the place where you bring your hogs to slaughter weight — should face south so the house or sleeping area opens away from prevailing winds. Many hog owners place the pen at the foot of the family garden, where garden wastes can be thrown over the fence to the hog or hogs. Some even maintain two garden sites and pen a shoat or two in the fallow site, where the animals naturally till and fertilize the ground.

Small Finishing Pens

I favor a small finishing unit made of native oak lumber and sheet metal with a capacity for two to four head; it can provide shelter in inclement weather, and it is easy to move if necessary. It keeps the hogs off the ground, which eliminates or reduces mud and helps keep parasites in check. Basically, it is made up of two parts: a small house and a slatted, floored pen fronting it. Some call the latter part of this unit a "pig patio" or "sunporch."

A pair of crossbred pigs is lolling in the autumn sun and growing toward a good harvest weight.

Width. Width is determined by the sleeping-bed option chosen. You can buy three-sided hog huts of sheet metal or sheet metal and wood construction in dimensions from 4.5 × 6 feet to 5 × 7 feet for about $150 new. If used, they can be had very cheaply; in the Midwest, good used huts often sell for less than half that $150 figure. These huts are light enough and of the right size to form these feeding units. I have seen the all-wood 6 × 8-foot one-sow farrowing units placed on long runners where larger numbers of hogs may be fed out. They will add to weight and cost but can accommodate up to six or eight head.

The foundation. Begin with two or three treated 4 × 6-inch runners of 12 to 16 feet in length. For a unit 4.5 to 5 feet wide, two runners should be adequate; for one that extends to 6 or 7 feet in width, three will be needed. Elevate the runners at the corners and center points with two to four concrete blocks and at a height just above the top of the ground.

You don't want them to sink into the mud and freeze down there. The runners will form the foundation for the entire unit; elevating them on the blocks will facilitate cleaning the pen and make it easier to load and unload the hogs.

The floor. Place a solid floor at least 2 inches thick on the half of the unit where the house or sleeping bed will be positioned. I prefer 2-inch full-cut native hardwood lumber, such as white oak. It is durable and, unlike some treated wood products, will not be a skin irritant to lighter-colored hogs. Solid sheathing such as plywood may form a surface too slippery for hogs to stand and walk.

The pig patio. Make the outside pen or patio floor from 2-inch-thick hardwood planking, with a 1-inch space between each board. The slots allow wastes to work through the floor and away from the hogs. At 1 inch, however, they will not catch the feet or legs of young pigs nor encumber larger hogs. The pen's sides can be made of 1-inch-thick

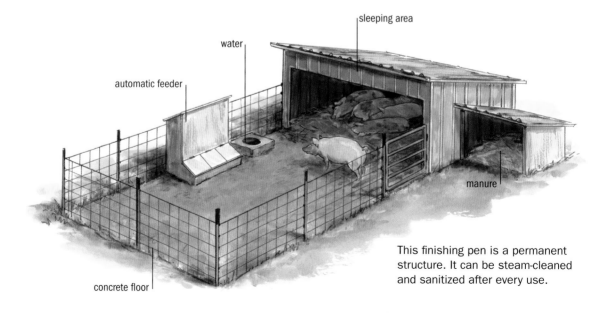

This finishing pen is a permanent structure. It can be steam-cleaned and sanitized after every use.

INSULATE THE ROOF

You may choose to use plywood or sheet metal and 2-inch framing lumber to build sleeping beds. Just be sure to insulate the roof. This will prevent condensation from forming on the inside of the roof and falling on the pigs and their bedding.

planking or 34-inch-high hog panels trimmed to fit the pen's sides and end.

Provide a 2-inch space between the bottom plank or panel bottom and pen floor to facilitate cleaning. In cold weather, the base of the unit can be insulated with bales of straw or sheet metal to prevent chilling drafts from coming up through the floor. Do this only around the floor; leave the slots open.

Advantages of a Small Finishing Unit

With normal care and maintenance, a small finishing unit can last 15 years or longer; can be moved about easily by a tractor equipped with a simple, two-pronged bale carrier; will not appear on property tax lists as a permanent structure; and will keep the hogs clean and easily accessible. This type of unit makes it possible to fit a couple of growing hogs into a space as small as 60 square feet, while keeping down the mud and maintaining the hogs comfortably.

Keeping Pigs on the Ground

Some owners prefer to keep their hogs on the ground. If this is your choice, you'll need to provide at least 150 square feet of pen space per shoat to keep mud problems from developing. In very wet or low areas, that amount of space may have to be tripled.

Hog lots become muddy not from the rooting activity of the pigs, which can be controlled with the use of humane nose rings, but from excessive foot traffic and those sharp pointed hooves. In drylots, feeders and waterers should be placed on concrete pads or hardwood platforms to prevent hogs from slipping and mudholes from forming around such well-used sites.

You'll still need a house for shelter, such as the small unit described previously, since it will be the animals' only dry retreat in wet or raw weather. It should have a step up to help keep muck out of the bedding area. Of course, you'll eliminate the pig patio, since you'll be raising your hogs on the ground.

Houses in lots or pastures still should have flooring atop the runners or, if floorless, be placed atop a 4-inch-high pad of stone, packed earth, or cull lime (lime not good enough for fertilization). Blocking between the runner ends with planking and shallow trenching around the range houses will help to carry runoff water away from the hogs' sleeping area. In raw weather, a 4-inch-deep layer of clean straw will increase the comfort level of the sleeping area by at least 10°F (5.5°C).

MANAGING FLIES AND WASTE BREAKDOWN

A couple of adult ducks of any variety can be an effective, natural method of fly control. Some hog owners also introduce red worms under their units to hasten the breakdown of wastes, for future use in the garden or wherever else composted wastes are needed.

To make the animals comfortable in this movable shed, a few drainage holes can be drilled in the flooring and covered with straw.

Numerous variations on these housing themes are possible, but remember that your primary goals are to keep sleeping hogs dry, protect them from drafts, and get them up out of the muck. I have seen hogs housed in everything from junked cars to houses made from old pallets and shipping crates. And they were doing just fine, thank you.

Fencing Options

To contain your growing hogs within a lot, use electric fencing, steel panels, or wooden gates. These fencing choices make it simpler to take down and rotate lots or to enclose idle garden plots. With electric fencing, use two charged strands; one should be 4 inches above the ground; the other, 12 inches.

Hog panels are all metal, can be moved easily by one person, and attach quickly to steel posts. They are 34 inches high by 16 feet long. New, they sell for $18 to $25 each, but they sell used for about half that much. At one recent sale where very little in the way of other livestock equipment was offered, I bought five of these panels for just $20 total. To erect pens for growing and finishing hogs from such panels, 5-foot-long steel posts are adequate.

I still see a few rolls of 26-inch-high woven wire around. If the gauge is heavy enough (top-of-the-line wire), it can be rolled and unrolled repeatedly to make swine pens. Its height was originally set so that it could be unrolled between rows of field corn to contain hogs set there to glean or "hog down" the corn crop.

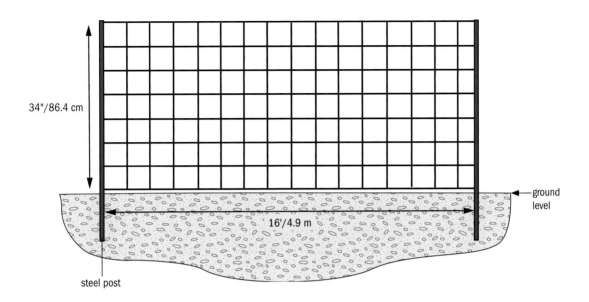

34"/86.4 cm

ground
level

16'/4.9 m

steel post

Hog panels are all metal, can be moved easily by one person, and attach
quickly to steel posts, as seen here.

Essential Equipment

Besides facilities to house hogs and fencing
to keep them in, you need a few other essential items that will help you easily and safely
handle hogs, even if you start with only one.
The few simple pieces of equipment discussed
below are all that you really need to comfortably and securely house one or more growing/
finishing hogs. The feeling of comfort these
materials provide for the hogs will promote
rapid and efficient hog growth. They will also
ensure the safety of both the animals and the
producer.

Put your equipment in place on the farm
before you own any hogs. This lets you avail
yourself of any pig-buying opportunities that
might arise. It will also help prevent you from

rushing into expensive expenditures for equipment due to spur-of-the-moment pig buying.

Loading Chutes

A simple loading chute 22 to 30 inches wide
and long enough to extend from the ground or
pen floor to the truck bed can be made from
little more than scrap lumber. Such chutes
generally run from 3 to 9 feet in length. They
should be narrow enough so that hogs cannot
turn around inside them but wide enough
that someone can walk behind the hogs while
loading.

Solid sides will make the hogs feel more
secure in the chute, and they will be easier
to handle. Solid sides also prevent shadows,
shifting light patterns, and outside activities

from upsetting the hogs. Cleats in the floor 8 to 12 inches apart will give them more secure footing. The chute decking should be made of material at least 2 inches thick. A few holes or small slots in the floor decking will allow rain and urine to run through, speed floor drying, and protect it from rot.

A bit of straw may encourage a hog to climb up an unfamiliar chute, but be sure to clean the chute floor after every use to extend its life. Inspect it often for raised hardware or broken boards that can injure the hogs or cause the chute to fail. Drop a hog through a rotted chute floor and you may then have to carry it onto the truck to get it loaded. Storing a chute indoors will certainly prolong its life, but if storage space is at a premium, it will suffice to give the chute a good cleaning after each use and set it out of the muck.

The best way to keep hogs moving up a chute is to walk behind them slowly, blocking any retreat with a short gate or hurdle (see illustration on page 53).

Handling Panels

Invest in an assortment of short, solidly made wooden gates and hurdles. **Hurdle** is an old term for a short, solid gate formed from 4 × 4-foot sheets of plywood, often called pig boards. Hogs can't see through them, and they are useful for sorting, loading, and restraining the animals.

A loading chute should be 22 to 30 inches wide — narrow enough to keep hogs from turning around but wide enough for someone to walk behind while loading them. Although it's the most expensive option, the diamond pattern gives the pigs extra traction. At Willow Valley, we use much less expensive wood floors with 1×2-inch cleats.

When snugly pressed between a barn wall or gate and one of these handling panels, a hog can be held safely for brief health treatments, sorting, or evaluation and measurement with a weight tape. These short gates are also useful for matching a loading chute mouth to various truck and trailer racks to provide safer and smoother loading and unloading.

You may have seen poles for handling hogs, but these are seldom used today. It's harder to direct hogs with poles than with gates or hurdles, and it's easier to become injured.

Form a hurdle from 4 × 4-foot sheets of plywood. Hogs can't see through them, and they are useful for sorting, loading, and restraining.

HANDLE WITH CARE

Just as with any other large animals, when handling hogs in close quarters, plan ahead and leave yourself at least two ways out. The hog has a totally undeserved fearsome reputation, but its size does dictate thoughtfulness and due consideration when you are working in close proximity to it.

Move about your hogs with a quiet and assured manner. Use an electric prod *only in most extreme instances*, such as when a sow refuses to get to her feet after farrowing. Never shout, and never try to work them when angry or stressed yourself. Never back them into a corner — they then have no choice but to strike out by lunging past you or biting. As my wife reminds me, you are supposed to be the one with the bigger brain in those situations.

When handling young pigs and shoats of up to 70 pounds, there is no substitute for the hands-on approach. Handy "grab points" are the ears and hind legs. That tail makes an awfully tempting handle, but you should never attempt to lift a hog of any size by the tail. It is a part of the spinal system.

Nevertheless, as a last resort, I have tailed as well as eared many a shoat into the back of a pickup. And at weigh-in time for our local 4-H hog show, I've seen youngsters apply more grips and holds than I'm apt to encounter in watching a whole year of professional wrestling on TV.

Over short distances, I will carry a pig by the hind leg and in a head-down position. If I carry it by the hind leg nearest me, the pig seems to travel more easily and more quietly.

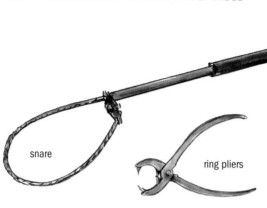

snare

ring pliers

Restraint Devices

A **snare** or set of **gripping tongs** ($20 to $30 new) is also good for restraining hogs of market weight and even larger. The snare surrounds the hog's snout; the tongs apply pressure to the back of the neck. Both are pressure points and restrain the hog in such a way that you can administer any necessary health-care treatments.

A restraining device, either gripping tongs or a head gate, must be used when a ring is placed to prevent rooting activities. Ring pliers clip a ring across the end of the nose or end of the gristly tissue around the nose. (See Protecting Pastures on page 158 for more on ringing.)

Feeding and Watering Equipment

Feeding and watering equipment can be as elaborate or as simple as your taste — or pocketbook — allows. The two essentials are keeping the drinking water clean and fresh and keeping the feedstuffs palatable so none are wasted.

Homemade containers. Here at Willow Valley, we still use a lot of troughs made from old steel water-heater tanks cut in half lengthwise. I weld bits of scrap iron to the ends to form legs for stability, and I weld bars across

the tops to prevent the hogs from lying in them and thus wallowing out feed or water. We also use a number of feed and water pans made by sawing the ends off 30-gallon plastic barrels.

For a great many years, we relied on a homemade feeder of about a 10-bushel capacity made with 2 × 4s, plywood, sheet metal, and a bit of found hardware. It was as ugly as homemade soap, but it wore like iron.

Store-bought containers. We have a number of rubberized pans and tubs and a couple of one-hole self-feeders suitable for one animal or a small group of two to ten hogs. The pans and tubs are inexpensive and easy to move, and in cold weather they can be flipped over and stomped on, to pop out ice.

You can buy a one- or two-hole self-feeder of sheet-metal construction at farm-supply stores for a bit more than $100; it will hold 1 to 5 bushels of feed. We owned a few made from 2 × 6 inch lumber that lasted for many years when bolted solidly to a pen gate a few inches above the pen floor. The life span of these feeders hinges on your efforts to maintain and keep them clean.

Fountain-style waterers. There are a number of fountain-type drinkers that can be attached to 30- and 55-gallon drums to provide drinking water in volume during warm weather. They can be set into a pen corner and will hold enough drinking water for several days. Used, these fountains run $10 to $20 each; new, they are in the $50 to $60 range. They are durable and quite simple in design; I once owned one that I kept in service for more than 10 years.

A simple waterer that will work for one or two shoats and takes up little pen space can be made from a 4- to 5-foot-long piece of 6- or 8-inch polyvinyl chloride (PVC) pipe, one pipe cap, and a nipple-type or Lixit waterer head. Cap one end of the pipe, then affix the nipple through a low hole on that capped end. The nipple must be a gravity-flow model rather than a pressure type. When all of the gluing is thoroughly dry, the tube waterer can be wired into a pen corner at the height appropriate for the pigs you are holding. This unit will hold a good amount of drinking water, is quite inexpensive, and can be easily stored when not in use.

Preventing Feed and Water Waste

The flow on a self-feeder should be tightened down, so that a hog has to work a bit to get all the feed it wants. Setting it to flow too freely can result in wasted feed. If you see feed on the ground around the feeder, you have a waste problem: As much as 10 percent of the feed you offer to your hogs can be lost through simple waste. I place only a week's worth of feed into a self-feeder, to ensure that it stays fresh and palatable.

If a hog is allowed all the water it can drink in 20 minutes twice a day, its water needs will be met. Still, water is the most important of all of the animal's intake. Troughs will keep water available to the hogs around the clock in all but freezing weather. Bars or rebar rods welded across the tops of the troughs will keep hogs from lying in them and wallowing out the contents.

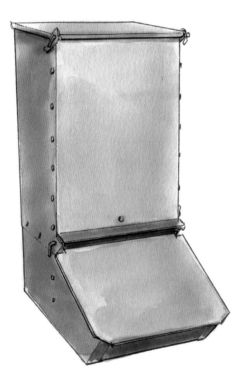

A one-hole self-feeder is suitable for one animal or a small group of hogs.

PROVIDE ADEQUATE TROUGH SPACE

Provide at least 12 inches of trough space for each hog in the pen to accommodate the animals' herd instinct to eat all at the same time. One space or hole in a self-feeder can accommodate three to five head of growing/finishing hogs because it keeps the feed before the hogs at all times. A water trough should be large enough to contain at least 12 hours' worth of drinking water.

These young shoats are drinking from a freeze-proof automatic fountain. The balls move up and down, allowing the hogs to drink.

This water barrel with two drinkers, or founts, could be positioned in a fenceline to provide drinking water for two small groups of hogs.

We have some deep wooden troughs put together with screws and caulked seams that we extend a short distance into our hog pens beneath a gate panel. Simple wooden legs support the portion of the trough that extends outside the pen. The hogs cannot wallow out any of the water, the troughs are easy to fill from outside the pen, and the drinking water in them stays much cleaner and more palatable.

Buying a Pig or Two

The old nursery rhyme goes, "To market, to market, to buy a fat pig. Home again, home again, rig-a-jig-jig." Well, buying a pig or two may not be quite as easy as this rhyme implies. Still, it is a fairly simple, straightforward task.

A meat hog of good type will yield about 85 percent of its liveweight (about 60 percent lean yield) in various cuts and meat products. Unfortunately, hogs are not all ham and chops, but one or two hogs fed out each year will go a long way toward meeting the protein needs of a typical family of four. If fed out at least two at a time the hogs will be more content, because they are herd animals. The second animal can always be sold or used by other family members or friends if your family doesn't need it. If sold, the money will help offset some of the costs of the animals and feed.

Where to Buy?

The best place to buy feeder pigs is at their farm of origin. There you can often view the **sire** (father), **dam** (mother), and siblings, which will give you an idea of what you can expect from the pigs as they grow out and reach finished weight.

Pigs at an auction are often stressed from the transport and handling. There is also a very real risk that they might have been exposed to disease organisms or sick pigs.

ESSENTIAL EQUIPMENT FOR RAISING HOGS

- Housing unit and pig patio or ground space and housing unit
- Fencing
- Loading chute
- Handling panels
- Snares or gripping tongs
- Waterer

When to Buy

Early spring may be the most expensive time of the year to buy feeder shoats because fewer sows farrow in the cold months of December, January, and February, when the early feeders are born. Still, these pigs are desirable, because they will reach a good slaughter weight before the weather grows excessively hot and humid.

Late summer through fall is also a good finishing period, as the pigs usually grow out before the late-autumn rains and winter cold begin, when feed consumption is impaired by temperature and weather extremes. These are the classic pigs for the winter's meat.

Barrows or Gilts?

While they may grow a bit slower than barrows, I favor feeding out gilts. Not only do they hang a leaner carcass, but they can also be pushed harder with a hotter (higher-protein) ration. In other words, growth can be accelerated if the cost of richer feedstuffs is no object. A **hot ration** might have a protein content of 15 to 16 percent or even 17 percent.

Barrows tend to pack on a bit more finish in the latter stages of the feeding period. In many

If fed out at least two at a time, feeder pigs will be more content because they are herd animals.

parts of the world, intact male animals are fed out for slaughter to take advantage of naturally occurring male hormones, which seem to enhance growth rates. However, in those regions hogs are also slaughtered at a lighter weight and younger age, before any trace of a **boar taint** can develop. Some folks describe boar taint as a strong cooking odor and a "slightly off" taste.

Cost

Expect to pay a bit more per head for a single pig or two than the going rate for pigs in groups (**droves**) because the farmer needs to be compensated for the added bother and stress of working a whole group to sell just a couple of pigs. Farmers also may find that by reducing the size of the group to be sold, the value of the remaining pigs is lessened.

Most commercial finishers want to buy single-source pigs in groups as large as possible, to keep down handling stress. The larger the group, the more the hogs generally bring per head.

A couple of rough rules of thumb: A 40-pound feeder pig should sell for between 1.75 and 2.25 times the going per-pound rate for butcher hogs; a 40-pound pig should cost as much as the going price for 100 pounds of butcher hog. Add a bit more if you are going to buy just one or two head. Heavier pigs should cost correspondingly less per pound, and pigs at 25 to 30 pounds a bit more per pound.

TIPS FOR BUYING FEEDER PIGS

Here are a few guidelines to help you with feeder-pig selection:

- The greatest variety and highest quality of pigs are available in the 40- to 60-pound weight range.
- If a pig is lighter than 40 pounds, it is a young pig and more prone to stress and setbacks. It may not hold up to the rigors of life in the finishing pen or be old enough to cope with marketing and transport.
- If a pig is much heavier than 70 pounds, it should raise suspicions that the animal is slow growing or stunted and over age for the weight.
- Crossbred pigs are generally more vigorous and faster growing than purebreds.
- Pigs should be crosses from a planned breeding program and not simple mongrels.
- Buy pigs that have been castrated and healed if you are seeking barrows.
- Buy pigs that have been treated recently for both internal and external parasites.
- Buy pigs that are well past the stress and strain of weaning.
- Bear in mind that while gilts may grow more slowly than barrows, they will generally produce leaner carcasses and can be pushed to grow faster with more nutrient-dense rations.

Feeder animal prices are very much affected by feedstuff prices. As grain prices go down, the demand and prices for feeder animals increase. When grain prices go up, there is less demand for feeder stock. Feed costs and feeder stock prices generally tend to move away from each other in opposing directions. Long term, however, the best cure for low prices is those same low prices, since they dry up excessive output.

Evaluating Feeder Pigs

A good pig for feeding is not simply a scaled-down model of a finished hog. It should show indications of good muscling and potential growth. Muscling is demonstrated by roundness or flaring in the areas where the primary pork cuts are to be found: the ham and the loin.

Growth potential can be seen in free and easy movement and a large, "stretchy" frame. Frame size is evident in the width between front and back legs, bone diameter as demonstrated in the legs, and width and depth of side.

In making final selections, many old hands focus on fine details, such as width between the eyes, size of the jaw, or even foot size. Where these dimensions are of larger-than-normal size, the reasoning is that the rest of the animal will "grow to them." The width between the eyes is one I rely on, as it does seem to be borne out by continued width throughout the body cavity.

When approached, a pen of pigs should rise up and move to the far side of the pen away from the intruder. They will then be lured back by their own curiosity to check out this

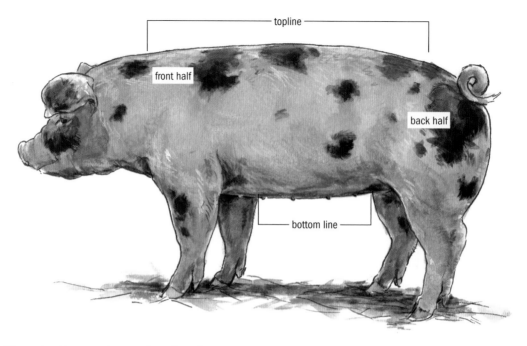

When checking out pigs, analyze them by parts: the feet and legs, the topline, the bottom line, the front half, and the back half.

PIGS TO AVOID

Not buying someone else's problems simply means, "Don't buy any sick or sorry pigs" (see chapter 10). Here are some easily detected indications of bad pig health and in some cases, the probable causes. These signs should raise a red flag immediately:

- **Exceptionally long and coarse haircoats** and heads that appear too large for the pigs' bodies appear on animals that are generally overaged, stale, and badly stunted. "Unstunting" a pig is no simple task and is seldom cost-effective.

- **Twisted, swollen, or misshapen snouts** are indications of advanced **atrophic rhinitis**, a disease that robs greatly from growth and performance and that can lead to death (see chapter 10). Do not buy even one pig from groups that contain these kinds of pigs.

- **Sniffling sounds as the pigs rise up and move about** may be a symptom of respiratory ills.

- **Discharge from the nose and eyes** or slight bleeding from the nose can indicate rhinitis.

- **Nodes or swellings (called knots) in the jawline** and other places are likely to be abscesses. They sometimes can be lanced and drained, but such procedures can introduce the abscess-causing organisms to your premises. Steer clear!

new feature in their space. If they appear to act sluggish, they generally are ill or lame.

A feeder pig is an animal generally weighing 40 to 70 pounds that is sold to a farmer/feeder to be fed out (raised and fed) to market weight. Such pigs are generally between 8 and 12 weeks of age.

Beginning the Selection Process

The pig-selection process should begin with your approach to the pigpen. Focus all of your senses on the pen as you close in on it. Listen for labored breathing, sneezing, coughing, or other sounds of respiratory distress. Certain diseases, such as transmissible gastroenteritis (TGE), an infectious gastrointestinal disease, create fecal material with a very distinctive odor (more about TGE appears in chapter 10). A cloud of dust or excess mud can indicate a stressful environment that might impair pig health or performance.

In selecting pigs, it is nearly always best to go with first impressions. Even the very best animals can be picked apart if you go over them hair by hair.

Form a mental image of what you value and the traits you want to find in a pig. Then compare the pigs being driven before you to that image of the ideal pig.

For faster analysis, break the live pigs into five overlapping zones: (a) feet and legs, (b) topline, (c) bottom line, (d) front half, and (e) back half. I will go into more detail about these different areas a bit later in chapter 8, but they do help in considering whether or not a good pig is the sum of good parts.

Note closely the environment in which the pigs have been raised, then try to buy animals from farm environments that most closely match what you have back home. This will do much to reduce stress on the newly arriving pig or pigs.

- **Dull, sunken eyes or a listless manner** are both indicators of an emerging health problem.

- **Dull haircoat, hair on end** (a sign that the body is trying to trap heat), or an excessively dirty haircoat, are signs of stress. Pigs with hair on end and piled closely together are obviously chilled and undergoing severe stress. If you note a greasy appearance or a spotty haircoat, or an excessive amount of scratching, look behind the ears for lice or lice eggs attached to individual hairs.

- **Swelling in foot and leg joints** can indicate injury or arthritis. Pigs presented in a ring or pen with little or no bedding generally have good feet and legs. Be on guard where straw bedding is piled deeply, however, as it can conceal much about foot and leg conditions.

Keep in mind that dramatically obvious ills such as ruptures — hernias that are sometimes called "busts" — and broken or downed ears are reparable but at considerable cost. Ruptures may be an inheritable trait. And don't take a pig with a stiff gait, front legs that are too straight, hind legs that are tucked too far under the body, knocked knees, front legs that both appear to be coming out of the same hole, or hooves with toe points of uneven size. These are indications of inferior body type, poor growth, and proneness to respiratory ills.

A lot of feeder animals are now being grown out in **controlled-environment nurseries** (temperature regulated) that hold the pigs constantly indoors in small pens. These pigs cannot always make the rapid adjustments needed for movement outside or to simpler, sometimes chillier housing. Pigs are normally best moved from such units when they weigh 60 to 70 pounds; they should never be moved in times or seasons of extreme weather. In very cold weather, they will pile up like cordwood to keep warm and may even smother or crush each other or develop rectal prolapses.

Listen and watch closely as the pigs rise up and lie down. These are the times when feet, legs, and underlines are often the most viewable. They are also the times when the pigs are most often apt to cough, sneeze, or experience labored breathing due to respiratory ills.

Don't buy someone else's mistakes, accidents, or **tailenders** (the slow growers in a peer group). Every pig has a history, and until you're satisfied that you know all there is to know about it, keep your hands at your sides and your wallet in your pocket. I have a friend who actually sits on his hands at most livestock auctions, just to be sure.

Signs of a Healthy Pig

A healthy pig will display an alert and curious manner. Any pig you buy should also have these signs of vigor and good health:

- **A bright and clean haircoat.** To me, this is a must in feeder-pig selection.
- **Free and easy movement.** This is vital if the pig is to remain sound and perform well while on feed.
- **Good growth for its age.** Simply selecting from the largest pigs in a pen

or group can do much to build a selection program with an emphasis on individual performance.

Feeding Hogs

On the subject of swine feeding, I am reminded of an incident that happened a fair number of years ago. On a warm fall afternoon, I sold a fellow about a dozen head of 40-pounders at a good price. After loading them, we sat down on the tailgate of his pickup for one of my favorite parts of hog production, the check exchange and a chance to shoot the breeze.

His first question was, "Now what do I feed them?" It was almost enough to cause me to let go of the check and unload that set of shoats. Almost. However, I decided to use his question as an opportunity to share some hard-earned experiences on the subject of feeding hogs. Above all else, remember that the reason to feed hogs or any other livestock species is not to save money but to make money.

The old rule of thumb was that in growing from 40 pounds to a handy market weight (230 to 240 pounds), a hog would roughly consume 10 bushels (roughly 685 to 750 pounds) of corn, and 125 pounds of protein supplement. The math still holds up pretty well, although many producers now feed rations that have higher crude-protein levels or are more complex in their formulation.

Swine rations, though seeming to grow ever more complex, are not available as widely and in as great a variety as they were even a few years ago. Not long ago, our local grain elevator carried three complete sow rations in the bag and would grind and mix several more to order. Some suppliers carry only a single growing ration, and at others you may have to put in a special order.

Be on constant guard for freshness; be concerned if bags are stained or dusty or the contents are caked, have a bad smell, show mold, or are excessively dusty in their composition. Many elevators still continue to offer special lot orders in amounts of 300 or 500 pounds. With small numbers to feed, it is nearly always the best practice to plug into your local supplier's complete feed program, which should be readily available in easy-to-handle 40- or 50-pound bags.

Specialty and Not-So-Special Feeds

Widely seen now are a number of specialty growing rations for show pigs. These have high protein levels, special ingredients that often change, and contents that many natural-based producers may not wish to use in their programs. Read those feed tags, folks, and make sure you fully understand them. Ask lots of questions, and bear in mind now that livestock feeds are being sold in many places where there is no one on staff with the ability to answer those questions.

With the reemergence of localized food production, many producers are striving to put a local stamp on their feeding practices. For example, I have been corresponding with a producer in New York who is feeding out his heirloom hogs in part by having them glean in an apple orchard and feeding them some of the by-products of apple processing. These practices will certainly affect the flavor of the pork produced and make it very attractive to today's consumers. I suggested he even offer pork items that were cured and smoked with the wood from old and damaged apple trees.

Many others now are growing out hogs on pasture or in woodlots and are seeking to feed locally available by-products from gardening or food processing or otherwise adding certain extras to the ration. A number, for example, are boosting mineral content with kelp and related products.

On range, or when adding to the ration food items with a lesser-known nutrient content, it is best to keep the hogs on a complete dietary ration appropriate for their age and role as sow, finisher, grower, and the like. Consider your additions as a little something extra and offer them in modest amounts that the animals will clean up quickly. Hogs are like big kids and can easily render their diets completely unbalanced by choosing items they favor in taste over less palatable, healthier others.

The substance left over from the production of ethanol is getting a lot of recent media play as a by-product used for feeding livestock. Used only where it can be simply and inexpensively transported, it is essentially brewers' grain—a by-product of ethanol made from dent corn that has been used for decades, mostly in cattle rations.

Touted for use in all sorts of livestock rations and in far higher levels than in the past, this grain by-product is, I believe, just one more rather thin justification for ethanol production. The classic nutrition texts call for no more than 10 percent brewers' grain in the rations of gestating sows and older growing hogs. Now, some sources are boosting this figure to 30 percent, and the same sources have okayed its use in the rations for other classes of swine. I wish I had a better feeling about this.

One of the best feeding investments anyone can make is an older, unabridged copy of Frank Morrison's *Feeds and Feeding*. This great old book contains all of the basics of livestock feeding and feedstuffs' composition. And it was put together back in the day when the mixed-stock farmer was king and regionalized farming practices were the norm.

Meat Scraps and Animal Protein Supplements

For a great many years, meat scraps and **tankage** — by-products of the packing trade — were dried and processed into an easily handled form. Meat scraps and skim milk were once about the only protein supplements fed to hogs, along with open-pollinated corn that was much higher in crude protein than today's heat-dried hybrids. Together these three items created simple, nutrient-dense rations for all classes of hogs and pigs weaned at around 8 weeks of age.

I buy into that old-time argument that it takes meat to make meat. However, the threat of **mad cow disease**, spread by the consumption of meat scraps from the spine and brain of prion-infected animals, has certainly changed the thinking of many on this, and I have certainly never advocated the feeding of kindred species' meat scrap. The addition of meat scraps, tankage, blood meal, and bone meal have been all but abandoned in food for hoofed stock and poultry, but they are high-protein and nutrient-dense feed items that are not easily replaced.

Some sources of animal protein, generally fish and dairy, can still be found in some of the more protein- and nutrient-dense, complex rations such as pig **starters** and early-stage growers. A couple of the more natural programs — humane, chemical-free, and complex in formula — are now calling for early-age rations supplemented with whey or egg-based products. If the animal-based protein source is from a species totally distinct from the species being fed, the chances of any potential health problems transferring are greatly reduced, in my opinion. I support the adding of fish, dairy, and eggs to rations, although some consumer groups prefer those that are vegan. It is widely known that in the wild, hogs eat a lot of red meat.

There are now a number of totally vegetable-based rations for hogs at nearly every stage of life, but bear in mind that hogs are true omnivores in their natural state. I believe the use of animal-based proteins as feedstuffs will continue; this is simply too large and valuable a resource to be turned into landfill. Even some gardeners now are refusing to use bloodmeal and bonemeal as soil amendments.

If the use of animal protein resumes, however, the hog industry will have to get a far better handle on how to safely process and handle the "waste," and today's producers have to be responsive to the legitimate concerns of their customers.

Feeding Corn

Swine rations concocted between 1900 and 1940 were simpler than most seen now — many hogs were finished on pasture or while gleaning grain fields. Still, as noted above, rations were filled with nutrient-dense ingredients. Not only was the tankage supplement 8 to 10 percentage points richer in crude protein than today's widely used protein supplements based on soybean oil meal, but the corn in use then was also more nutritionally dense.

In the good old days, skim milk or good legume-pasture and corn varieties like Reid's Yellow Dent or Bloody Butcher were all anyone needed for late-stage finishing or sow maintenance. Those old, open-pollinated field-corn varieties often tested in the 13 to 16 percent crude-protein range. This was far better than the 8 or 9 percent levels assigned to modern hybrids, and many hog producers are assigning a value of just 6 percent when formulating rations with heavily heat-dried corn.

Nevertheless, yellow corn continues to be the basic feed grain for hogs and most other livestock species. Some form or variety grows in every one of the 50 states; the national yearly crop is measured in billions of bushels, making it both widely available and modestly priced in most years, and it can be used in a number of forms for swine feeding. Many producers still feed some ear corn, sometimes even allowing the hogs to glean it in the field, and others feed a great deal of shelled corn, especially if feeding on the ground or in open troughs to better counter waste.

A once-common practice was to keep barrels of shelled corn soaking in water to improve palatability and digestibility when fed. More common now is the practice of feeding corn in a meal form after running it through a hammer or burr mill of some sort. Some of these mills are also combined with a mixing tank that blends all the ingredients of a ration to give it the bite-after-bite consistency that many hog raisers now favor. Grinding the corn does improve its digestibility, and the hog is then better able to utilize the nutrients it contains.

Soybean Products

If yellow corn is the foremost feed grain in use, its counterpart in protein supplements is **soybean oil meal**, commonly abbreviated SBOM. It is a meal-type by-product of the manufacture of soybean oil. Varying in content from 44 to 48 percent crude protein, it is blended with corn or other feed grains, then supplemented with vitamins and minerals to create

FEEDING ALTERNATIVES TO CORN

Alternatives to corn as the primary grain in swine rations include grain sorghum or milo, barley, and wheat. For best results, all should be ground to a meal form and should probably replace no more than 50 percent of the corn in the ration mix. The so-called bird-proof grain sorghums should not be used, because the factors that deter birds from feeding on it as it ripens can also greatly reduce its palatability.

The newer yellow and white strains of grain sorghum are even better suited for use in swine rations, because they contain more vitamins than the red varieties of milo. Wheat may be an even better crude-protein source than corn, but it is generally a more costly grain to feed; if ground too fine, it becomes gummy when chewed, which can reduce overall feed consumption.

Barley and milo have 90 to 95 percent of the feeding value of No. 2 yellow corn. Supplementation may be needed in the form of soybean oil meal or a **premix** pack. Premix packs vary in content and are generally used to add vitamins, minerals, and flavorings to boost nutrient density and increase consumption. For example, most rations containing milo are boosted with what is commonly termed an "A-D-E pack." It contains substantial amounts of the vitamins A, D, and E.

a complete growing/finishing ration. It is also often the base protein supplement in young-pig and breeding-stock rations.

Soybean meal can also be found in two other forms. The first is commonly termed **hog forty**, as it has a 40 percent crude-protein content and is already supplemented with the more commonly required vitamins and minerals to form a complete swine ration. Mixed as four parts grain to one part hog forty, it creates a good growing/finishing ration with a 15 percent crude-protein content. This 4:1 mix is known in the Midwest as the classic basic stock mix.

The other form is **extruded soybean meal**. This is made by running whole soybeans through an extrusion screw that processes them with both heat and pressure. Temperatures in the screw may rise to 300°F (149°C), and the resulting oily meal typically has a crude-protein content of 36 to 38 percent; however, it does have several more percentage points of fat than does regular soybean meal. This fat is a valuable energy source, increases ration palatability, and may even reduce some of the dust associated with many grind-and-mix rations.

There are alternatives to soybean oil meal and meat scraps, including some dairy-based products, but they tend to be higher in price. The alternatives contain more complex and easily digestible proteins. They do create very good, very palatable rations, and they are a must in formulating rations for very young pigs.

Using Dietary Additives

A number of dietary **additives** for swine rations offer goals as varied as boosting nutrient levels, increasing consumption, and improving digestibility. More of us in the hog industry are now mixing in any one of a number of regional products that contain fish by-products, yeasts, and kelp meal, to reduce performance setbacks during such times of stress as farrowing, weaning, and when hogs are moved. The one we happen to use at Willow Valley is Immunoboost, which is made in Texas.

In the past, I have experimented with various probiotic products and top dressings for rations. **Probiotics** are starters and enzymes that activate or enhance the activity of the flora in the gut. Diarrhea can strip the gut of its nutrient-grabbing **villi** and "good" bugs, as can antibiotics. Probiotics are supposed to help restore equilibrium. While some of these probiotic products have performed very well, I believe that fresh ingredients blended correctly are the real essentials of a good ration.

A practice favored by a number of old hands called for use of at least 30 to 50 pounds of meat scraps in place of some of the soybean oil meal in a ton of swine feed. This gives the animals multiple sources of protein — just as they would obtain while foraging for themselves in the wild. A lot of these same veteran producers also grind and mix a bale of good alfalfa hay into every ton of complete feed they formulate. This is an added nutrient source and also gives the ration some extra bulk.

Choosing Types of Rations

Answering the question, "What type of feed?" sounds almost too simple; just put what you want them to have in front of them and they will eat it, right? Well, not really.

Early in the last century, growing hogs were given X amount of corn each day. The corn was counted out to provide so many ears of corn per head or so many scoopfuls of corn per pen of hogs. For many months each year, the hogs were on legume-rich pastures with a bit of corn fed each day. Skim milk or tankage might be offered in a trough once or twice each day.

But swine feeding progressed. In its next advancement, hogs were offered access to corn — generally shelled corn — at all times, and offered protein supplements once a day in a trough or some sort of self-feeder. If all went well, the hogs were free to balance their own rations and grow at a good pace. The trouble was that the protein feeds were generally more palatable than the feed grain; they were consumed quickly and in ways that knocked rations completely out of balance. Not only did performance suffer, but hogs fell ill with gastric upsets. A better method of feeding hogs was found with the grind-and-mix method.

The grind-and-mix method. The next step in the progression of hog feeding entailed grinding the corn into a meal to increase its digestibility; as much as 5 percent of whole-grain corn fed to a hog can pass through its system totally undigested. The grinding process breaks down the kernels' seed coats and makes the nutrients contained within far easier to digest. The ground corn is then mixed with the other ration ingredients to form a "complete feed."

These grind-and-mix rations remain in widespread use. They can be manufactured on the farm with a tractor-powered implement called a grinder/mixer or with a self-contained unit commonly called a stationary mill. Such rations can also be bought in bag or bulk form from many feed dealers and elevators.

The real plus with grind-and-mix rations is their bite-after-bite consistency of nutrient content and texture or form. With all in order, every pig that gets to the feeder gets a fully balanced, complete ration. It is a very good approach to swine feeding as long as particle size does not become too fine. Very fine feed particles can cause problems of palatability and in some hogs can even trigger ulcers. Grinding and mixing also can add to feeding costs. If bought in bulk, however — there is a 1,000- to 2,000-pound (unbagged) minimum purchase at most elevators — it is one of the least costly forms in which to buy complete swine rations.

Pellets and cubes. The other commonly available form in which to buy swine rations is as pellets or cubes. Under heat and pressure and with a binder such as alfalfa, the feedstuffs are formed into pellets or cubes that range from ⅜ inch long up to about the size of a man's thumb.

There are two real pluses with this process:

- It may free as much as 5 percent more of the nutrients contained within the feedstuffs (a result of the application of both heat and pressure to them).
- When fed in troughs or on the ground, pellets and cubes create less waste than meal-type rations. This is, however, the most costly form of ration processing.

Still, processing costs are a small consideration for the producer with but one or two hogs on feed. The ease with which pelleted feeds can be handled and fed, their widespread availability (in some areas they may be the only option for producers buying swine feed by the 50-pound bag), and the potential for increased efficiency make pellets or cubes a top feed choice for owners with small numbers of hogs.

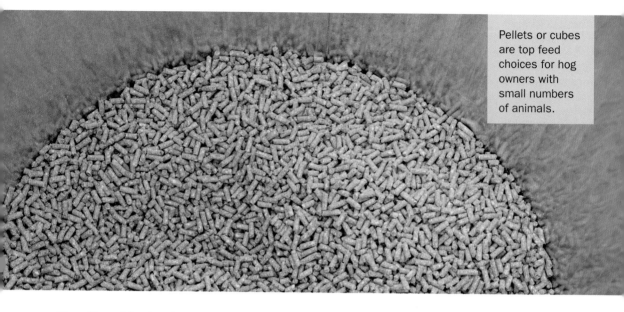

Pellets or cubes are top feed choices for hog owners with small numbers of animals.

Feeding Methods

There are two approaches to feeding out a pig or pigs to market weight.

Free choice or **full feeding** is the first of these approaches. Simply put, the hogs have access to all the feed they might want 24 hours a day. By the numbers, a hog on full feed will eat about 3 percent of its weight daily in feedstuff. By the end of the feeding period, a finishing hog will be eating 7 pounds or a bit more of feed daily.

Limit-feeding a hog to about 90 percent of appetite, the second approach to feeding out an animal, will produce a slightly trimmer hog and will somewhat reduce daily feed costs. It will also extend an animal's time on feed; as a result, you will probably realize no overall savings on feed costs. Still, this is an option if you wish to produce some extra-trim pork for the family table.

As a finishing hog ages and grows, its growth rate and feed efficiency slow, and much of the weight gain in the late stages of the growout period is often finish (fat cover) rather than lean or muscle gain. A traditional feeding practice to help the producer cope with this natural pattern has been to reduce the crude-protein content of the ration as the hog grows.

Today's leaner hogs, and especially the better gilts, can be pushed to grow faster with richer feedstuffs. I keep my growing/finishing hogs on a 15 percent crude-protein ration from 40 pounds straight through to end weight. I

TRADITIONAL GROWING/FINISHING HOG-FEEDING PRACTICE	
Weight	**Ration of Crude Protein**
From 40 to 75 or even 125 pounds	15 to 16 percent— a true grower ration
From 75 to about 175 to 200 pounds	14 to 15 percent — a grower/finisher ration
From 175 pounds to desired weight	12 to 13 percent — a true finishing ration

even use this ration with the boars that I grow out to 300 to 350 pounds before offering them for sale. They continue to gain well and remain lean.

If you do opt to make changes in swine rations, do it gradually over a period of three to seven days. Sudden ration changes can cause gastric upsets and stress in hogs of all ages.

Buying Feed

The producer with two hundred head of hogs on feed has a far greater incentive to shop for feed bargains than the producer with just one or two head, but some comparison shopping should still be in order for the smallholder. Few other livestock feeds have remained in the lower price parameters that have bracketed swine grower/finisher rations. They move up and down with corn and soybean prices, but these are often computer formulated to better control price and may use by-products to do so.

By the 50-pound bagful, feed regularly falls into the 8- to 12-cents-per-pound range; one growing hog will generally consume 650 to 750 pounds of such feed as it grows from 40 pounds to slaughter weight. Still, I have seen price differences as great as 1 or 2 cents per pound between comparable complete feeds at suppliers in our immediate area. I don't advocate switching brands every time you buy feed, but price savings like those matter even if you own just one or two head — enough to justify spending a bit of time on the phone with your local feed suppliers.

Selecting a Feed Dealer

There are more considerations in feed-dealer selection than mere price. At some dealerships, small accounts are more appreciated and better served than at others. For example, when confronted with use and management questions, some feed dealerships can do little more than tell you to read the little paper tag sewn to one end of the feed sack.

I favor a small local elevator that seems to specialize in small accounts, carries a wide variety of feedstuffs, offers special services and delivery of amounts as small as 500 pounds, will special-order products, and carries a good line of animal-health products. While prices on many products may be a bit higher, these are more than offset by the special services and considerations offered for the smaller producer.

DAILY CONSUMPTION

To give you an idea of how much hogs eat, the following chart details the average daily consumption of our hogs on full feed.

Age	Weight	Daily Consumption	Percent Crude-Protein Feed
8 weeks	40 lbs.	1.2 lbs. pelleted starter-grower	17–18%
12 weeks	75 lbs.	2.5–3 lbs. grind and mix	11–15%
16 weeks	125 lbs.	4–5 lbs. grind and mix	15%
To market	150–230 lbs.	6–7 lbs. grind and mix	14–15%

Limit-feeding a hog to about 90 percent of appetite will produce a slightly trimmer hog and will somewhat reduce daily feed costs, but it will also extend an animal's time on feed. As a result, you will probably realize no overall savings on feed costs.

TOO MUCH MANURE? COMPOST!

Some years back, an Amish farmer on my feed sales route was considered quite remarkable by the staff of the Cooperative Extension Service. He had one of the first 70-bushel soft-wheat yields in our area; it was considered an astonishing amount of wheat.

Grudgingly, Extension staffers came out to ask the secret to his success. His answer did not help them sell either bigger tractors or more chemicals. He told them simply, "I had an open January, six big boys still at home, and three barns full of manure."

Where there are hogs eating healthfully, there's plenty of waste to go around, and hauling it from place to place soon becomes one of life's daily details. In fact, the first "love letter" I sent my wife-to-be was composed while mucking out hog houses.

Manure and spent bedding are two products of the swine venture for which it is difficult to assign a value. I know you can go into our fields each year and quickly tell where the last load of manure was spread. That crop has larger plants at every stage of development and is lush and greener.

We typically fork wastes directly onto a wagon, pickup, or spreader and take it straight to the field. In inclement weather or muddy periods, we toss it into small piles just outside hog houses and barns. With the spent bedding of sows and their litters, we add to the nearby piles until the pigs are moved to different quarters. After a few days, a cap forms on a pile and the plant and waste material begin breaking down under the cap and generate heat.

In the fall and winter, when we spread the composted material a few inches deep over cropland, it adds nutrients, boosts organic matter, and improves tilth. Because the straw bedding absorbs much urine, we are able to add it to the soil, too.

Feed Storage

Storage of feed for one or two pigs need not be at all elaborate. Most producers buy feed from one week to the next. This allows them to keep their feedstuffs fresh and to manage feed costs by averaging out price highs and lows throughout the entire feeding period.

Fifty pounds of feed can be dumped into a metal or plastic trash can with a tight-fitting lid. The feed will stay dry and protected from birds and vermin — the main causes of damage to and contamination of feedstuffs.

If you do find a good price on feedstuffs and opt to stock up on them, a 55-gallon plastic or metal drum with a secure-fitting lid will hold up to 350 pounds of ground or pelleted feed or shelled corn. Set the barrel on concrete blocks or an old pallet to keep its bottom dry, and place a rodent-control station safely beneath it.

For even larger amounts of feed, I use a pair of old chest-type freezers in which the compressor or motor has failed. These are often available for the hauling from repair shops and appliance dealerships, and some of the larger models will hold up to 1,000 pounds of feedstuffs. The latching mechanisms must be removed as a child safety measure, but the door will still seal in place. These freezers can serve as quite satisfactory, watertight minibins.

A metal or plastic trash can is a good way to store small amounts of feed. It should have a tight-fitting lid.

PROTECTING FEED

Feed for hogs must be protected from **mycotoxins**, which are produced by fungi. These can accumulate undetected in swine feed, resulting in all sorts of problems, ranging from impaired growth to a susceptibility to infectious disease. The USDA Animal and Plant Health Inspection Service (APHIS) cautions that young pigs especially are susceptible to the mycotoxins called aflatoxins, produced by aspergillus molds. Aflatoxins are often associated with drought-stressed corn and moisture levels of more than 14 percent. Fumonisins are another type of mycotoxin; at high levels they can cause rapid accumulation of fluid in the lungs.

The two other major mycotoxin threats to swine are vomitoxin, which reduces feed intake and weight gain, and zearalenone, which reduces or inhibits reproduction in females.

Some mycotoxin problems are things a producer can't do anything about, APHIS says. What you can do to protect your pigs, however, is place grains in dry storage and keep moisture down. Clean feed-storage bins often and use ground feed rapidly, so that mycotoxins don't have the chance to develop to threatening levels.

They protect the contents from moisture and vermin, and they can be positioned all around a farmstead, even at penside. If you set them on a single tier of 8-inch-high concrete blocks, you'll find that they are further moisture proofed and their contents are more accessible.

Bringing Your Pigs Home

With feed and facilities ready at home and pigs selected, the aspiring hog raiser is ready for the hands-on side of pork production. This begins with getting the animals safely home and off to a good start.

If you're hauling them in a pickup or stake-sided trailer, cover the front of the racks with a tarp or sheet of plywood to keep chilling drafts off the pigs while in transit. Bed the racks with straw to a depth of 4 inches in damp or cold weather. In hot weather, haul early in the morning or late in the day; use damp sand or sawdust for bedding.

At home, unload the pigs into a well-bedded sleeping area. It may be necessary to block the small animals away from a portion of the sleeping area, to discourage them from developing the habit of dunging inside the house. Wetting down a far corner of the pen for the first couple of days may encourage them to begin dunging in that area.

Managing Stress

Moving pigs stresses them, and one sign of stress is going off the feed. It is not at all uncommon for newly moved pigs to go off their feed within a few hours. This shouldn't last for more than several days to a week, however, and the pigs should continue to drink water even while off their feed.

To help animals through this stressful time, many producers add a vitamin/electrolyte

---

USING SELF-FEEDERS

If your pigs will be allowed free access to self-feeders, tie the feeder lids up or open for a few days, until you are sure that the pigs are large enough and able to operate them. The same holds true for lids or covers on the drinking fountains.

---

product to the pigs' drinking water. This helps them maintain both health and body condition should they go off feed or have trouble adjusting to a new ration. Adding flavored gelatin to drinking water will further increase water consumption and can also be used to mask any unpleasant tastes from medications that are administered through the water. (More on medications appears in chapter 10.)

Top-dressing feed with the same gelatin powder will draw the pigs to it and increase consumption there, too. Still, never forget that abundant drinking water is a hog's most important feedstuff.

Young, newly moved, and stressed pigs should not be offered anything in the way of new or exotic feedstuffs. Many producers try to buy some of the feed the pigs were previously eating or match it as closely as possible, to further help them through those first stressful days at their new location.

Parasite Control

Unless you have been assured otherwise — and have some sort of documentation — assume that all pigs have at least been exposed to all of the major internal parasites or worms, as well as to external parasites and any localized problem parasites. They will need to be dewormed (refer to chapter 10 for information on this).

Swine Finishing

Now that your hogs are settled in and you've got parasites in check, the tasks remaining in hog finishing may seem to amount to little more than keeping the feeder and waterer filled, but there's more — a good deal more.

Avoid Unnecessary Drugs

There seems to be an injection or in-feed antibiotic for every sneeze, sniffle, or hiccup a hog may have. When you are finishing hogs, however, your first job is to avoid the temptation to become drug and medication happy. If a finishing hog has been carefully selected, is well fed, and is kept dry and comfortable, it should easily cope with most changes in the weather and shifts in the season.

There are swine feeds available that do not contain any type of antibiotic, although at times it may seem difficult to find them. We have found that hogs not on a steady diet of antibiotics grow as well as any others and if a problem does emerge, they respond better and more quickly to a treatment-level dosage of a health product than do those hogs given a steady low-level or subtherapeutic diet of the in-feed version of the antibiotic.

Signs of Health Trouble

Read chapter 10 for detailed information on specific diseases. Suffice it to say here, however, that pigs that are slow getting off their beds, that are eating less, that stand up looking drawn, or that demonstrate any of the other signs of ill health need further examination and possibly treatment. They may have a bacterial problem or be coming down with an infectious disease. A rectal thermometer can also do much to help you diagnose health problems as they emerge. The normal body temperature

> ### HOGS AREN'T GARBAGE DISPOSALS!
>
> While on the subject of exotic feedstuffs, I should point out that hogs aren't four-legged garbage disposals. You can build pretty good hog rations from items as diverse as stale bakery goods and cannery wastes, but note that some very rigid laws now govern the processing and feeding of modern food wastes. Most states, for instance, require that food wastes be cooked to 180°F (82°C) in an approved facility before they can be given to livestock. These laws apply even to the farmer raising one or two hogs for the family table, although they may seldom be enforced on small farms.
>
> To build a truly balanced ration from ever-changing table scraps and garden wastes is nearly impossible. It is best to consider these extras, something over and above a pig's regular ration, and to never feed them in amounts so great as to disrupt the animal's digestive system.

of a hog is 101 to 102°F (38 to 39°C); a fever indicates that the body's systems are rising to its defense, whereas a below-normal temperature may indicate that body systems are shutting down, as occurs with kidney failure.

Note any health problems as soon as they start, and summon the veterinarian as quickly as possible. With nearly any health problem, treatment is quickest and least costly if begun in the earliest stages.

If you have just one or two pigs on hand, perhaps the best health appliance in which to invest is some means of restraint for the

BEWARE OF HOT WEATHER

With growing and finishing hogs, we have encountered more health problems in hot weather than in cold. The first hot spell of the summer, especially if it comes on quickly and is accompanied by high humidity, can take quite a toll on finishing hogs.

In such conditions, hogs have had no time to become acclimated to the weather and thus react poorly to it. When it's hot out, provide extra drinking water and keep it fresh and clean. You may even wish to sprinkle hogs with water from above in the hottest part of the day. These measures will add greatly to hog comfort and survival.

animal. Use of tongs, a snare, or heavy gates to push the animal against a wall will allow the veterinarian to do his or her work safely and quickly. (More on home medical supplies to keep on hand, as well as how to administer injections and treat wounds, appears in chapter 10.)

Finishing Weight

The last thing you'll have to determine when finishing hogs is what weight you want to feed them to. We personally prefer to feed our hogs for the freezer to just 220 to 230 pounds or even a bit less, but that's because we are only a two-person family.

Today's hogs can be taken to 250 to 260 pounds and still have quite lean and high-yielding carcasses. I have even seen some 300-pounders place quite well at market-hog shows — although at that weight, we're talking about a lot of pork and some large primal cuts, such as the hams. Most families no longer really want or can use whole hams.

A hog will normally yield about 70 percent of its liveweight in pork and pork products. We generally feed out two hogs a year for our needs: one for slaughter in late winter and another for the fall. We also find fresh and cured pork to be a most appreciated gift item during the holiday season, so we do a bit of our Christmas shopping from the hog pen.

With your hog fed out to the desired weight, your next step is the harvest, or slaughtering process. In the next chapter, I will discuss the various ways that you — the hog raiser — can become involved.

4

HOME BUTCHERING AND PORK PROCESSING

Slaughter and butcher are hard words for many people, and I concede that they are processes some of you may not want to take on as do-it-yourself projects.

These traditional farming activities can result in measurable cash savings, but they are also time-consuming and call for special skills. Here at Willow Valley, we have slaughtered and worked up as many as four hogs in a two-day period, but of late we find it quicker and simpler to have a butcher hog custom processed.

Our costs for custom processing run $12 to $25 for trucking, $15 for slaughtering, 32 cents per pound for processing (cutting up the carcass, wrapping the cuts, and fast freezing the meat), a small fee to render lard, and 35 cents per pound for special processing (curing, making into 4-ounce patties, making into links, etc.). A typical hog costs up to $90 to work up. If it is processed on a Monday, the fresh meat is ready by Friday or Saturday and the cured meat a couple of weeks later.

These prices have remained fairly constant in our area of rural Missouri for a number of years. Even here, new methods such as special seasoning for brat-type sausages push up the costs. East and West Coast prices will be higher, and the special handling required for organic product will cost more.

Slaughtering the Hog

In most of the descriptions I've read, butchering day dawns clear and cold, the smell of wood smoke wafts through the air, and geese fill the sky. For us, however, such days dawned when the bottom of the freezer started to show and the butcher hogs were at last big enough. In even further contradiction, our butchering day began in the late afternoon, when we drove the selected hogs away from the others to a pen near a scaffolding made of a tree line or a raised loader bucket on a tractor, erected for hanging the animals during skinning and cooling. There they were held off feed overnight to clear the gut ahead of dressing the carcass.

The next day, family and friends gathered at our farm for the butchering. We always had at least one good shot on hand and a couple of strong backs to work under the experienced and watchful eye of my late father-in-law, R. E. Perkins Sr.

To kill the hogs to be butchered, we used a .22 loaded with long rifle bullets; as a safety

---

Lean pork rolled and tied with butcher's twine for cooking

measure, only one person went to the pen to kill the selected hog.

For a killing target, draw an imaginary X from ears to eyes in your mind's eye. Step back a few feet, then shoot just off center from where the lines intersect. The hog needn't be restrained, as it is unaware of what is to happen. Death is instantaneous if the shot is true. Do not do this in the presence of other hogs, however, as they may become agitated.

With the hog down, two people quickly step in, grabbing up one front leg apiece and dragging the hog head first from the pen. Drag the animal with, rather than against, the haircoat, which is the best way to drag any downed animal. Then position the carcass on the ground with the head facing down a slope to facilitate blood flow and sever the jugular vein. Make this incision deep and long to be assured that it is free flowing. Bleeding can take up to about 10 minutes; there will be a gurgling sound, then reduction of flow before the process is complete.

Imagine an X drawn from ears to eyes. Shoot just off center where the lines intersect.

Severing the jugular for a rapid bleed-out of the carcass

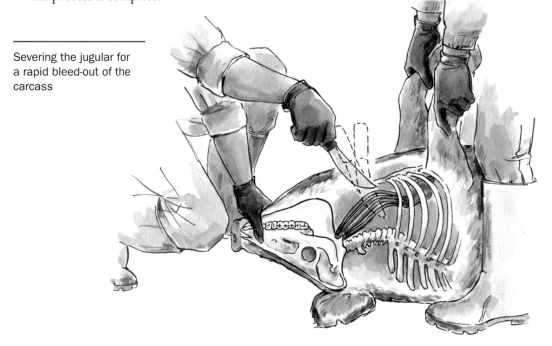

GENTLE HANDLING = TENDER PORK

The less handling and stress a hog goes through just prior to slaughter, the more tender the pork will be. If a hog is fearful and becomes stressed, chemical changes take place that affect the meat. To reduce stress before slaughter, some farmers pen waiting hogs away from the slaughter area, so they cannot glimpse or smell any of the slaughtering or butchering process.

Skinning

With the bleeding complete, begin the skinning process and move the carcass to the scaffolding, where you have affixed a pulley-type fence-wire stretcher to the top beam. You'll find that a tractor with a front-end loader is useful in transporting, elevating, and suspending the carcass. Attached to the stretcher is a gambrel of sorts (a frame used by a butcher for hanging carcasses) that can be used to attach the hog to the stretcher and hold the carcass open while you work on it. For a gambrel, we most often used a well-scrubbed singletree from my late father-in-law's collection of draft-horse rack. This is the linking device for attaching a horse harness to an implement. When used for butchering hogs, it spreads the animal's hind legs.

Place the hog beneath the main beam, its back legs adjacent to the now fully extended fence stretcher.

Beneath the scaffolding's top beam is the main work area. You may have another skinning setup, but be sure to pick a site free from any obstructions and one that has a safe surface upon which you can stand and work. We work hard to keep this spot free from blood, wastes, and general clutter by keeping tubs and buckets nearby for collection.

I've seen people go into this like Jim Bowie at the Alamo, but all you really need are a good pocketknife and an 8 × 10-inch butcher's knife.

You'll see all sorts of fancy, high-priced skinning knives advertised, but I've found the old reliable three-bladed stockperson's pocketknife to be more than adequate for this skinning task. We also keep at hand a sharpening tool or butcher's steel, along with a whetstone (dry hone).

With a bit of care, two people can work at the skinning task — and this certainly makes the work go more quickly. To skin the hog, follow the steps below:

1. With a sharpened pocketknife, make a short, vertical cut around the foot and just above the hoof inside each rear leg.

2. Work carefully to expose the tendon that extends down the back of each leg. It appears as a white cord, very strong and taut, and even a slight nick will cause it to sever, so it must be exposed very carefully.

3. Free each cord from the surrounding tissue.

4. Position the singletree between the hind legs, and attach the hooks on each end of the singletree beneath the exposed tendons. The hitch ring on the singletree is hooked onto, or wired to, the lowered end of the fence stretcher.

5. Raise the carcass from the ground to a good working height. Should a tendon

break or be severed, wrap heavy-gauge wire repeatedly around the foot and fasten it to a hook on one end of the singletree.

6. Begin skinning the carcass from circular cuts made just above each of the rear hooves.

7. From the cut ringing the foot, make another, much longer cut down the inside of each back leg up to the vent and then completely around it.

8. Remove the skin by working the knife blade between skin and muscle, pulling down on the skin as it is loosened.

9. Another circular incision just through the skin at the tailhead will enable you to completely skin the hams.

10. With the hams complete, make a long cut just through the skin from vent to head.

11. Working from this incision, continue cutting, scraping, and working down the hide by pulling on it as it is loosened. Use special care when skinning the flank area, because the skin is very thin there and must be worked back very carefully from the flank and belly.

12. As the work moves down the inverted carcass, the front legs are skinned out in a manner akin, but inverse, to the procedure for the back legs.

13. Make another cut completely around the head, and the hide should come away in one piece. At this point set the hide well away from the work area, to be disposed of or processed following the dressing chore.

14. Remove the head from the carcass with a series of twisting movements and careful cutting with a heavy-bladed knife or small hatchet. Set the head aside for separate skinning.

Skinning the head, while a bit time consuming, is simply a matter of removing the skin from the back of the head forward, generally in modest-sized segments. Take extra care when removing the skin around the jowls, where it is very thin, and when removing the fleshy end of the nose. The head can then be cooled out in a large container of freshly drawn water and used later in distinctive pork products, such as headcheese.

This is a carcass skinned halfway down. It can now be raised higher to give you easier access for the rest of the skinning process.

THE ADVANTAGES OF SKINNING

Scalding is a more traditional process than skinning. (Instructions for it appear beginning on page 84.) Here at Willow Valley, however, we find skinning is an easier and quicker process. Skinning is also the safer of the two options. It eliminates both the need to handle scalding hot water and the dangers of working around a scalding hot vat.

Evisceration

With the skin and head removed and the tail and all four feet in place, the carcass is ready for evisceration, or disembowelment. Keep in mind that all viscera should be handled carefully to prevent punctures, tears, or cuts, which could cause contamination of the carcass or organ meats.

1. Starting again at the vent area, make a long downward cut to the head to begin the opening of the body cavity.

2. Using a heavy-bladed knife or small hatchet, break open the pelvic girdle at the point where the hams naturally separate (often an assistant will aid with this task by pulling the back legs apart).

3. With the pelvic girdle opened, continue the cut the length of the body cavity, inserting the index and second fingers of your free hand into the cavity on each side of the knife blade to draw the skin away from the viscera while you make the long downward cut.

4. Use the heavy-bladed knife to cut through the breastbone.

5. If the hog is a barrow, cut away and discard the **pizzle** (penis) at this time. Should the urinary tract next to the pizzle be full, tie it shut with a piece of clean baling twine, to keep urine from contaminating the carcass. A similar problem may be encountered with the bowel at the anus and can be handled in the same manner.

6. Place a large tub beneath the carcass to catch the viscera as they are removed from the body cavity.

Splitting the sternum as part of the evisceration process

TOOLS FOR SKINNING AND EVISCERATION

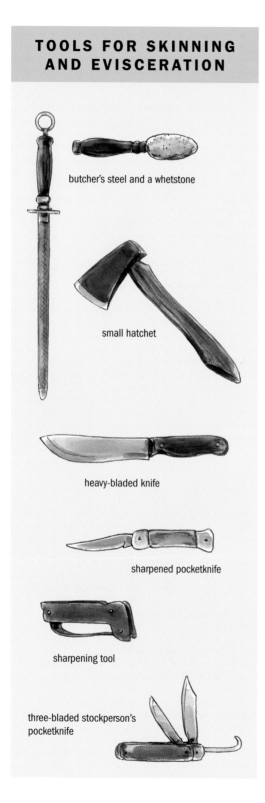

butcher's steel and a whetstone

small hatchet

heavy-bladed knife

sharpened pocketknife

sharpening tool

three-bladed stockperson's pocketknife

7. Begin the process of freeing the viscera from the anus and body cavity with the short blade of a pocketknife and the hands. Much of the viscera can be removed in a single large mass, although the lungs, heart, and esophagus will have to be removed after the chest cavity is opened.

8. Remove the liver carefully to protect both its quality and its future flavor. It is positioned adjacent to the gall bladder, and rough handling or a puncture could contaminate it with bilelike fluids.

9. Split open the heart, and rinse it until it is free from all blood.

10. Place the organ meats in a basin of cool water, and put that basin in the refrigerator until it is time to package, label, and freeze the meat.

11. Remove the tail by making a cut ringing the tailhead that extends through the back, then take it out as you would a plug. Some producers scrape the tail and use it as a seasoning meat in certain vegetable dishes.

We dispose of the viscera by burying them in a site well away from water sources and to a depth that will discourage digging by wild or domestic animals. The hide can certainly be processed for later hobby or craft work, but time always seems to be short at butchering, so I generally dispose of the hide by burying it. It can also be composted in a large pile of sawdust positioned away from water sources.

Viscera are removed from the body cavity to be deposited in a bag or bucket. Organ meats can be separated and the remainder buried or burned.

A meat saw or handsaw can be used to halve the carcass for easier handling.

Cooling

With the viscera removed, we rinse out the body cavity with 10 to 15 gallons of cool water, or a bit more, and then leave the whole carcass hanging to cool out overnight. Following tradition, many producers leave the butchering task for the coldest weather to hasten the cooling out of the carcass; a temperature drop of about 20°F (11.1°C) from day into night, however, is adequate for the process. We have butchered fairly late into the Missouri spring and as early as late September in the fall. Where insects may be a problem, the carcass can be wrapped in cheesecloth without impairing the cooling process. The carcass should be hung close enough to the house that there are no problems with predators.

Although a cooled carcass is easier to halve, you can facilitate cooling by halving the carcass lengthwise down the backbone. We've used a hand meat saw for the halving task but prefer a regular carpenter's handsaw, as the blade has less flex. A well-cooled carcass is also much easier to cut up and process.

BY-PRODUCTS

By-products of butchering include blood, feet, offal (the waste parts), and hides. While commercial slaughterers are equipped to salvage all of these items, the home processor is unlikely to be so equipped or able to sell them.

Hog feet do figure prominently in a great many recipes; they were a personal favorite of my wife's grandmother. Still, feet and offal, especially from several head of hogs, will pile up in a hurry, cluttering the work area and making footing and movement there more difficult. I generally trundle them away to be buried in a safe place or share them with neighbors who have the time to process them correctly.

Blood is an ingredient in several types of sausages and other food items, but the time and effort required to gather blood and keep it sanitary can also take you away from the skinning of the more valuable carcass. Blood is very difficult to keep clean while being gathered, and unless several animals are slaughtered at once, it is often difficult to gather in large enough amounts to be practical. **Chitterlings**, or small intestines, have the same handling and processing problems as blood, and their appeal is often quite limited.

Scalding

Although we prefer skinning — it's faster, easier, and safer — many hog producers still choose to dress hogs via the scalding method. It's the way some families have always dressed hogs, and they wish to continue the tradition.

Scalding takes more preparation time as well as more equipment and manpower than skinning. At least two strong people will be needed to move the hog or hogs in and out of the scalding bath. You also must bring a large amount of water to a boil and keep it that hot.

The scalding can be done in either a large (100-gallon) barrel or a scalding vat. The water must be heated to 150 to 155°F (66 to 68°C). Before scalding, the hog should be killed and bled as previously described.

To be scalded correctly, the hog must be completely immersed in the hot water. Two people using short log chains hooked about a lightweight carcass can lower it into a vat and manipulate it up and down in the water until the scald is complete. With a larger carcass, a boom or hoist is in order.

Before you remove the hog from the scalding water, make sure the scald is complete by checking to see that hair can be pulled away easily from each side of the carcass. To keep the scalding water hot when more than one hog is being dressed, many folks heat pieces of scrap metal in the scalding-tank fire and then drop them into the scalding water.

When the scalding is finished, move the hog from the water to a solidly constructed table for scraping. To remove the hair and scurf from the carcass, begin by scraping the legs, head, and belly. Keep the carcass wet with warm water throughout the process. When you have completed the scraping, you can skin it and then break down the carcass later.

Use a bell scraper to scrape the hair off the scalded hide.

Breaking Down the Carcass

Meat cutting is both a skill and a science. As a home processor, you certainly do not have to be as precise in your work as a retail butcher when reducing a carcass to its most usable-size portions. This often involves little more than following the natural seams within the muscling.

Pork is made up of muscle fibers, connective tissues, fat, juices, and water. Young animals will have tenderer muscling than older ones; physical activity also affects muscle quality. The tenderest cuts come from the loin region — the boneless cut from the pork loin is, in fact, called the tenderloin. Tougher cuts of meat have thicker muscle fibers and more connective tissues. They come from the muscles the animal uses the most.

Hogs have both a fat **cover** over the carcass and fat flecks within the meat, which make the meat tender, juicy, and flavorful when cooked. Without a certain amount of fat, pork and any other "red" meat would be largely unpalatable. Among swine breeds, the Duroc and Berkshire are most noted for the flavor and eating qualities of their pork, even when they are crossed. Fat flecks within the muscling in these breeds function in much the same way that marbling does in good beef. The Japanese are now especially partial to the pork from Berkshires and their first-generation crosses, and such pork brings a premium when sold in the Far East.

Immediately after cutting, meat has a distinctive purplish red color, but exposure to oxygen soon turns it to the bright red "meat" color with which we are most familiar. To break down a hog carcass, we use only a long-bladed boning knife, a heavy-spined butcher knife, and a meat cutter's handsaw. We send out our fresh hams and shoulders to have them sliced to the desired ⅝-inch thickness on a power handsaw for a small fee. For such slicing, partially frozen pork is easier and safer to work with than fresh pork at air temperature.

PRIMAL CUTS

Primal cuts are large cuts that are often transported to butcher shops for further butchering and sale. There are seven primal cuts in the halved hog carcass:

1. The leg, which is comparable to the round in beef and produces boneless leg, ham, and ham slices or steaks

2. The loin, which can produce blade chops, loin chops, butterfly chops, country-style ribs, back ribs, Canadian-style bacon, loin roasts, and tenderloin

3. The side pork, which yields the bacons

4. The spareribs, which yield both ribs and salt pork

5. The Boston shoulder or butt, from which can come pork cubes, Boston roasts, shoulder roll, and that Midwestern favorite, pork steaks

6. The picnic shoulder, which yields up roasts and steaks, ground pork, and sausage

7. The jowl, which can be cured for seasoning meat or sliced like bacon

No single carcass can produce all of the above, but the beauty of home processing is that you can give over as much of the carcass as possible to your family's favorite cuts.

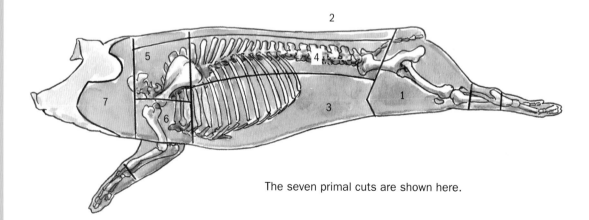

The seven primal cuts are shown here.

Butchering

As noted earlier, the butchering process is essentially the disassembling of the carcass at the joints and along naturally occurring seams within the muscling. Having the right equipment will help you obtain the cuts of meat you want.

Butchering Tables

I have seen butchering done on the kitchen table — and done a bit of it myself — but a more stable, easy-to-clean work surface is preferable. Workbench-type tables topped with a fiberglass laminate, or even stainless-steel restaurant tables bought secondhand, are excellent for supporting butcher work.

One of Midwestern furniture makers' earliest uses of walnut was in simple butchering tables. It is a sturdy wood that stands up well to the oils and greases common in the butchering process. Valuable though walnut furniture may be now, here in hog and corn country it was used initially for this most utilitarian of purposes.

A 10-foot-square area in a shop or outbuilding with smooth and easily washable floor, ceiling, and wall surfaces is the ideal site for a home butcher shop. It should be centered on a large worktable that can be accessed from both sides and both ends.

Cutting Tools

Quality tools kept sharp and clean do much to expedite and ease the butchering task. Cutlery for breaking down the primal cuts into table-ready portions includes a boning knife, a butcher's knife, a cleaver, and a meat saw (a short handsaw will do in a pinch). A butcher's steel and dry hone are necessary to keep a good working edge on the cutlery. Dull edges slow your work and increase the risk of injury. Cutting through hair or bone seems to take an added toll on cutlery edges.

MEAT-CUTTING TOOLS

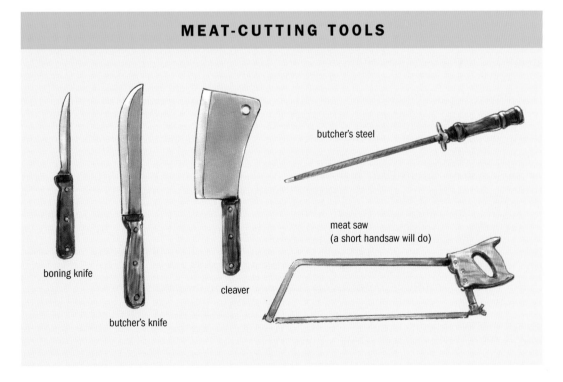

boning knife

butcher's knife

cleaver

butcher's steel

meat saw
(a short handsaw will do)

You will need a small meat grinder for turning scraps and trim into sausage or pork burger. A band saw with a meat-cutting blade will certainly simplify the process of slicing steaks and chops into the desired serving thicknesses.

From Primal Cuts to Chops and Hams

Separate the carcass into the larger, primal cuts with the butcher's knife (an 8- to 10-inch heavy-spined blade); use the cleaver where necessary to disjoint major carcass segments. Cutting completely around the joint points and sharply rotating the primal cuts to be detached will often hasten the joint's separation. As a safety measure during this work, wipe the cutlery handles and your hands often to ensure a solid grip on the tools at all times. There are gloves you can buy that have material on the palms to give you an especially good grip.

With the primal cuts separated, you must then decide how to further break them down for table use and long-term storage. For example, you can debone the loin to yield the tenderloin, which can then be sliced into medallions or butterflies, or you can leave the bone in and slice the loin into pork chops. The ham can be left whole, broken down into two or three large roasts, or sliced into ham steaks of various thicknesses. Fresh, uncured ham is often said to be "green," meaning it has a greenish pall.

Chilled meat is the easiest to work with; some folks even partially freeze pork that is to be ground into sausage or burger. This is a virtual must if the grinding is to be done with a small, hand-turned grinder. Steaks and chops are normally sliced to a thickness of 5/16 inch for fast and even cooking. Roasts are normally broken down into segments weighing 2 to 4 pounds — large enough to center a meal for the entire family.

Someone Else Could Do It

Perhaps at this point I should repeat that the costs of butchering, and your own personal tastes, may favor having your hog or hogs commercially processed. Commercial operators have the skinning and cutting tools, refrigeration for rapid chilling, carcass-handling implements, and manpower to do the job at a fraction of the toll it takes on many country households. When the butchering begins, virtually all effort from all available bodies must be given over to it until it is completed.

Think of it in these terms: A 210-pound butcher hog (quite small by today's standards) will still have a dressed weight of about 150 pounds and a packaged or wrapped weight in the area of 120 pounds. At home, those 75-pound (or heavier) carcass halves must be carried to the work area and handled largely by just one person. The meat must be kept clean and wholesome and moved to proper long-term storage as rapidly as possible. Work surfaces and utensils must be wiped down frequently, and you should wash your hands just as frequently.

Clearly, butchering even a single hog can fill up the family schedule for the better part of two days, and once started, it is a task from which you cannot withdraw.

On the other hand, even if you wish to add some personal touches to your meat, you can do so with commercial processing: simply request the return of a green ham or some bacon from the processor for home curing, if those are the cuts you enjoy.

Putting Away the Pork

When you're up to your elbows in the task of working up a carcass, you'll soon see the truth in the words of the legendary livestock nutritionist Frank B. Morrison in the ninth edition of his classic text, *Feeds and Feeding Abridged*: "Pigs exceed all other farm animals in the efficiency with which they convert feed into edible meat. They require much less feed and much less total digestible nutrients for each pound of gain in liveweight than do other farm animals. They also yield a higher percentage of dressed carcass, a larger percentage of the carcass is edible, and pork is higher in energy content than other meat."

A half of the aforementioned hog with its 75-pound hanging weight should yield 14 pounds of fresh ham, 12 pounds of loin or chops, 6 pounds of trim for sausage, a 12-pound bacon, 3 pounds of spareribs, a variable amount of lard (up to about 10 pounds), 11 pounds of shoulder for steaks or roasts, and 5 pounds of bone and shrink. Many of these weights include the bone, however, and thus are not 100 percent edible meat. Still, there is very little waste with modern hogs. I well recall sending a gilt to be processed and actually having to buy a few pounds of fat to add to the trim to make a sausage that would have enough fat for proper cooking.

What, then, are you to do with all this protein largesse? How can you put it away for a rainy day — or a BLT on an August afternoon when those tomatoes are truly ruby ripe and fresh from the garden?

Pork products can be frozen, canned, cured, or smoked. Freezing is now the most common means of home meat storage, and it is also perhaps the simplest and least time consuming.

HOW TO PRESERVE QUALITY

To preserve their quality, you should freeze pork and other meats as quickly as possible. Set the freezer control to 0°F (–18°C) or slightly below; higher freezing temperatures will cause larger ice crystals to form in the meat. Those crystals break down meat cells and fibers and adversely affect juiciness and texture in the meat when it is then prepared for the table. Rapid freezing, on the other hand, causes smaller ice crystals to form in the meat and thus preserves better eating quality. When the meat is completely frozen through, you can return your freezer to a higher temperature setting.

Freezing

The deep freeze or freezer is not a wonder appliance into which you simply drop things and find that they remain fresh and appealing for forever and a day. Understanding a freezer's limits will prevent potential illness and substantial financial loss.

Freezing simply slows the changes that affect food quality. Bacteria are not killed by freezing; they are simply halted from multiplying. How the meat is prepared for freezing, how the freezer is maintained, and how foods are thawed prior to preparation will all affect eating quality and wholesomeness. Freezer management is truly an ongoing chore.

One cubic foot of freezer space holds 35 to 40 pounds of food. Allow 4 to 6 cubic feet of freezer space for each family member; it will sustain them for six to nine months. A well-filled freezer operates better and more efficiently than one only partially filled.

You should attempt to freeze only those amounts of meat that will freeze thoroughly in 24 hours. Never try to freeze more than $\frac{1}{15}$ of the freezer's capacity at a time. A good guideline is to freeze only 2 to 3 pounds of meat for each cubic foot of freezer space.

To save space in the freezer, debone the cuts of pork as much as possible, and trim them into pieces as uniform and compact as possible, before wrapping them and packing them into the freezer. Because frozen meat is best when consumed as soon as possible following thawing, it should be wrapped in portion-size amounts that can be used at one time or for one meal.

Wrapping pork for freezing. All materials used for freezer wrapping should be both vapor- and moisture-proof. Heavyweight aluminum foil or freezer-grade plastic wrap is the best choice for wrapping bulky or irregularly shaped cuts. When you're packaging steaks, chops, or patties, place a double thickness of waxed paper between them for easier separation after they're frozen. Once you have properly labeled them, spread the new packages out in the freezer as much as possible to hasten their thorough freezing.

The basic closure technique for freezer wraps is the three-step drugstore fold, which forms an airtight seal that will protect the contents from freezer burn:

Step 1. Place the pork cut in the center of the wrapping material.
Step 2. Bring the two horizontal ends together and fold over until they are tight against the meat.
Step 3. Tightly fold one end and then the other, turn each end underneath, secure tightly with freezer tape, and label.

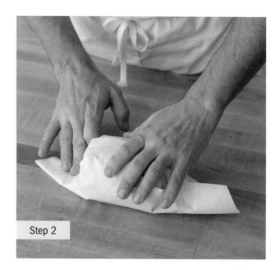

Step 2

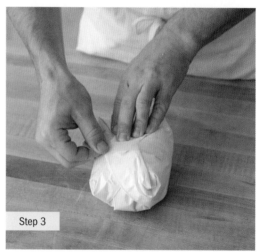

Step 3

Fold wrapping material tightly against the meat, as shown, and secure with freezer tape.

After wrapping, label the freezer packages with a grease pencil, permanent ink marker, or on a tape-type label. The label should note the cut contained within and the date it went into the freezer. Those pieces of meat that have been in the freezer the longest should be placed nearest the door for soonest use.

TIPS FOR FREEZING AND THAWING PORK

Although freezing hastens color changes in red meat, pork that is properly handled and wrapped — and then used within the approved storage times — should suffer little effect to its eye appeal, flavor, and juiciness. Do not, however, season ground pork before freezing. Most seasonings — especially sage — are intensified by the freezing process. Lard, after complete cooling, can also be stored in the home freezer. Freezing times that will best maintain the table qualities of various pork items are:

Cured bacon	1 month or less
Cured ham	1–2 months
Sausage	1–2 months
Fresh pork chops	3–4 months
Organ meats	3–4 months
Pork roasts	4–8 months

The best way to thaw frozen pork is in the refrigerator. For small roasts and steaks, allow 3 hours of thawing time per pound; for large roasts, 4 to 5 hours per pound.

A metal hand-turned grinder is sufficient for processing small amounts of meat.

Grinding the Pork

With the carcass of a hog broken down and cut up, you will have a good-sized pile of lean trimmings from all over the carcass for sausage. Before grinding, all animal heat must be gone from the meat; chill it, to make it firm. Cut the chilled meat into cubes or strips for easiest feeding into the grinder.

An electric grinder will certainly speed up the pork-grinding chore, but a hand-turned grinder is adequate for small amounts of well-chilled or partially frozen trimmings. All-metal grinders offer the greatest durability and ease of cleaning. They will either bolt to a large board or table or clamp to the edge of the work surface. They come with grinding plates of different sizes that you can use to double-grind the meat to the desired texture.

For the best taste, some fat should be ground with the lean meat to enhance the juiciness and flavor of the sausage. The fat content of pork sausage normally ranges from 20 to 30 percent; the latter is typical of many supermarket sausage products. We have some of our Willow Valley hogs custom processed to produce an 80 percent lean whole-hog sausage, which we sell at $2.50 to $4.00 per pound at our local farmers' market. The whole-hog sausage sold in most supermarkets is generally made with pork from cull sows rather than from younger, pricier butcher hogs.

LARD RENDERING

Another by-product of hog processing is fat for lard making — but only if the hog has been scalded, because skinning removes too much fat. Lard is a useful cooking product and the secret ingredient in the crusts of more blue ribbon–winning pies than you might imagine. And with a kettle of bubbling lard and a stringer of fish, you'll soon have an old-fashioned fish fry on your hands.

The fish will take a bit of doing, but the steps for rendering lard are fairly simple. Here are the instructions, along with some tips.

1. Remove all of the skin and lean from the backfat and all other fat trimmings. Fat from around the internal organs will yield a darker lard; it should be rendered separately, since it is usually discarded.

2. Cut the fat into small but uniformly sized pieces (1- to 2-inch cubes).

3. During the rendering, the fat should remain at around 212°F (100°C). Never allow it to rise above 255°F (124°C), because scorching can occur. If you are rendering in a kettle over an open fire, keep the fire low, and stir the rendering fat often. Do not render the lard in a copper or brass kettle; these metals can cause it to become rancid.

4. Skim off cracklings as they rise to the surface and turn brown. Hot cracklings can be pressed in a lard press to produce more lard. Cracklings are also a pretty fair snack food if you were raised country. They can be added to your favorite corn bread recipe to make cracklin' bread, and they will keep your hounds fat and slick like nothing else.

5. When all the water in the lard is evaporated, the rendering process is complete, and the heating should be stopped.

6. Strain the lard through several layers or thicknesses of cheesecloth into pails, crocks, or other storage containers.

7. Cool the lard immediately at a temperature near freezing. (To set properly, it must chill through.)

8. As the lard cools, stir frequently to achieve the desired creamy appearance. This will also prevent the lard from taking on a grainy texture. Dark-colored lard either was scorched or had too much lean meat to remain attached to the fat.

9. Air and light are harmful to lard, so containers should be filled to the top, sealed tightly, and stored in a cool, dark place. Package lard for freezing in amounts that you can use quickly following thawing.

Pork for sausage should first be run through the coarse blade of the meat grinder. Then mix it thoroughly with your hands, and spread it out thinly. Season it evenly, and mix it thoroughly again by hand. Finally, regrind it through a finer plate (I recommend the ⅛-inch-hole plate). Sausage to be frozen is best left unseasoned, and sausage for canning should receive no sage seasoning.

Preserving Pork

Drying, smoking, and curing are the oldest known methods for preserving meat. They yield a distinctive flavor, a taste of history some might say, and no wonder — even the great early Roman, Cato, was known to have had a favorite recipe for salting down hams. The practice of drying and then smoking meat can be traced all the way back to the ancient Egyptians and Sumerians.

Pork-preserving methods are time consuming by today's standards, and the finished products may require refrigeration for safest storage. Still, their taste, aroma, and texture can be produced in no other way.

Most cures are quite often closely guarded family secrets and may include such widely varied special ingredients as red pepper and cloves, but the two basics of most cures are sugar and noniodized salt. Salt is both a good preservative and a flavor enhancer. It moves through the meat by that process we all read about way back in high school: osmosis. It also helps inhibit bacterial growth. Sugar helps counter some of the salt's harsh edge, brings out further flavor in the meat being cured, and lowers the pH of the curing solution.

The third ingredient common to many pork cures is **saltpeter** (nitrate), which is different from curing salt. It fixes the desired taste

PORK BURGERS

One of the newer pork products is commonly termed "pork burger." It is an 80 percent lean, ground-pork product seasoned with one of a number of various seasoning mixes. The seasoning is blended in at the rate of ½ ounce per pound of ground pork. Pork burger is very good on the barbecue grill, and it makes one of the richest-tasting cheeseburgers I've ever had. It can be stored for up to four months in the home freezer.

and color in cured meat and protects it from the often-fatal botulism organisms. The nitrate question is a touchy one for many — nitrate has been associated with some forms of cancer, but it is the best protection the home processor has against botulism. If used, saltpeter should be stored and handled with great care and added only in the exact amounts set down in whatever curing recipe or formula you are using. Some people delete the saltpeter and make sure to use the cured product quickly.

Meat that has been frozen and thawed should never be used for curing. Ice crystals damage meat texture, so meat that has been frozen will be more vulnerable to spoilage.

The best pork for curing comes from young slaughter hogs in the 180- to 240-pound weight range. Hams for curing should be from hogs that have been scalded rather than dressed out by the skinning process because the skin (rind) is necessary to maintain juiciness and texture. I have no way to substantiate this, but it is my belief that the pork from hogs fed whole-grain corn and grown out more slowly has better table and curing qualities.

Curing Hams

The two most common types of cure are the **dry cure** and the **pickle cure**. In the dry cure, ingredients are rubbed into the surfaces of the meat. With the pickle cure, a brine-type solution is injected into the meat down to the bone with a large needle and syringe.

Dry-Cured/Sweet Pickle–Cured Ham

This combination cure is from the University of Missouri.

A pumped pickle cure is included because the quicker the cure reaches the center of the ham, the less risk there is of meat spoilage. A special culinary needle and syringe, which can be purchased from a store that sells cooking supplies and utensils, is the recommended way to apply the pickle cure.

7 pounds meat-curing salt

3 pounds white or brown sugar

1 gallon water

1 5-inch-long culinary needle attached to a plunger-type syringe

1. Thoroughly mix the salt and sugar together.
2. To prepare the pickle cure, dissolve 2 pounds of the salt-and-sugar mixture in the gallon of water.
3. Pump no more of the pickle-cure solution than is equal to 10 percent of the ham's weight.
4. Rub the remaining dry cure into the surface of the ham at the rate of ½ ounce to 1 pound of ham.
5. Allow the ham to cure on a table at 34 to 45°F (1 to 7°C) for 14 days.

Where I come from, the two words "country cured" are enough to set most mouths to watering in anticipation, and few delicacies are more eagerly sought out than true country-cured hams. Here in Missouri, a number of family farmers have gone into the business of producing country-cured hams and now ship them worldwide as gourmet treats. One of the most competitive events at many county fairs, as well as the summer-climaxing Missouri State Fair, is the cured-ham competition.

Generally, the "country-curing method" begins in December or January when the cold weather has settled in for a good, long spell. You'll want to begin with hams chilled to 40°F (4°C) that come from hogs slaughtered within the last 24 to 30 hours; this cure must be applied within 48 hours of slaughter. We are talking truly fresh here, and one of the high points of what is traditionally a week or more of butchering and related activities.

Few delicacies are more eagerly sought out than true country-cured hams.

Country-Cured Ham

Trim your ham of excess fat, to give it the traditional ham shape, while being careful to expose as little of its lean portion as possible.

2 pounds noniodized salt

1 pound white or brown sugar

1 ounce saltpeter (optional here)

1. Add the salt, sugar, and saltpeter, if using, to a large bowl and mix together thoroughly. To further enhance flavor and aroma, you can add black pepper, red pepper, or cloves to taste. This recipe should provide enough curing mixture for two hams, the production of one hog.

2. Rub the cure into all surfaces of each ham at the rate of 1¼ ounces of cure per pound of ham.

3. In the single-application method, all the cure is applied to the meat at once, with the lean areas receiving most of the mixture. See that the hock receives plenty of the cure as well and that it is applied uniformly to the face of the ham.

4. Wrap the ham completely in nonwaxed paper, place in a cloth bag, and hang shank-down in a well-ventilated, dry place. Allow 2½ days of curing time for each pound of ham; if the ham should freeze during the curing period, add 1 extra day for each day the ham might have been frozen.

5. When the curing time is up, unwrap the ham, remove the excess cure, rewrap the ham, set it to age in a cool (40 to 50°F [4 to 10°C]) place, and then check it weekly for signs of possible spoilage.

In the two-application method, initially apply ¾ ounce of cure per pound of ham. Wrap the ham, and place it on a shelf in a cool, dry place for 7 days. Apply the remaining ¼ ounce of cure per pound of ham, then handle exactly as above.

The Missouri-Cured Ham

*From my late father-in-law, R. E. Perkins Sr.,
comes this Missouri cure for the hams from a
200-pound butcher hog — a weight that has to
be considered prime for the family table.*

2 cups meat-curing salt
1 cup brown sugar
2 tablespoons black pepper
1 teaspoon red pepper

1. Combine the salt, sugar, black pepper, and red pepper in a bowl. Rub the mixture thoroughly into all surfaces of the ham.
2. Wrap the ham completely in newsprint.
3. Tie the wrapped ham with a string in a 1-inch-square, crisscross pattern.
4. Place the ham joint-down in a muslin sack.
5. Hang it in a cool, dry place, and let it drip. If the ham doesn't start to drip right away, take it down and examine it carefully for problems such as mold or insects.
6. Let the ham hang and cure for 6 to 8 weeks.
7. When the curing time is up, take the ham down and wash off the remaining cure. The ham can then be used right away, refrigerated, or frozen.

This is a very old recipe. If you are concerned about the inks and processing methods used in producing modern newsprint, consider using paper printed with the new soy-based inks.

Smoking

On the old homestead, building a smokehouse was third on the list of priorities, after building the house and barn.

That fondness for the rich, smoky flavor that hardwood smoke imparts to just about anything that runs, swims, or flies has continued to this very day. Hardwood smoke gives meat products a rich, full flavor and improves their keeping qualities. The wisps and tendrils of smoke that curled around the buffalo and pronghorn meat hung from the support poles of Native American tepees now flows through electric smokers, which are equally at home on small-town back porches and big-city penthouse terraces.

It does take a bit of our time to properly tend the smoke chamber, and matching hardwoods to create the desired color and flavor of the finished, fully smoked meat products is both art and science. My wife's great-aunt, Nova Warnka, relates an old-time measure to counter insect problems: a bit of sassafras wood added to the smoke generator will act as a natural insect repellent.

Country-cured hams should be unwrapped prior to smoking. Any excess cure or mold growth should be removed with a stiff brush and a rinse in cold water. Smokehouse temperatures

AVOID RESINOUS WOODS

Resinous woods such as pine, cedar, or other evergreens must never be used in meat smoking — or even in smokehouse construction, for that matter — because they can impart poor flavor to the meat.

for ham should not exceed 90°F (32°C)—what is commonly called a "cold smoke." The meat should be smoked for about 48 hours or until the desired rich amber color is attained. After smoking, the ham can be rewrapped and hung up for more aging.

Smoking mild-cured bacon. Follow the five steps below to create a flavor few will be able to resist:

1. Wash the cured bacon in warm water.

2. Hang the bacon in the smokehouse, and allow it to dry for between 48 and 72 hours, with the smokehouse door open.

3. When the bacon is dry, smoke it using your choice of hardwoods to fuel the generator. Use a cool smoke (under 100°F [38°C]) for 36 to 48 hours.

4. Smoked bacon is perishable, so following the smoking process, it should be refrigerated or frozen.

5. Any rind should be removed from the cured bacon prior to freezing.

Sliced bacon can be frozen for two to three months, but for longer storage wrap and freeze the bacon in chunk-sized pieces that can be used quickly after thawing.

Curing Bacon

As popular and tasty as cured hams are, they quite often have to take a backseat to that center of the great American breakfast, crisp bacon. Bacon is a breakfast staple from Vermont to the high Sierras. And along with mild pork sausage, it is one of the products we direct-market from our small swine herd.

Mild-Flavored Bacon Cure

This recipe comes from the University of Missouri.

Use only the freshest of pork bellies, cooled to 42°F (6°C), and begin the processing within 24 to 30 hours following slaughter. The cure must be applied within 48 hours of slaughter. The green pork bellies can come from skinned or scalded carcasses. Be sure not to stack warm bellies one atop the other, as that practice interferes with airflow, drying, and sanitation. Trim the bellies carefully.

    2   pounds meat-curing salt
    4   pounds white or brown sugar
    3   ounces saltpeter

1. Place the salt, sugar, and saltpeter in a bowl, and mix thoroughly.
2. Apply the cure by rubbing into the bacon surface ½ ounce of cure for every 1 pound of green belly.
3. Cure the bellies in a well-ventilated room or outbuilding and on a tilted table, so that the moisture produced by and during the curing process will drain away from the bacon.
4. Pile the bacons in a crisscross manner on the table to a height of no more than 4 bacons.
5. Allow the bacon to cure for 7 days. If it should freeze during the curing period, add 1 extra day for each day it might have been frozen. The bacon can then be smoked.

There's Pork in This? — Pork Recipes

Even though we farmers producing truly flavorful, delightfully textured pork poke fun at the marketing slogan, "Pork — the other white meat," that ad has probably reached just as far around the world as the names "Coke" and "Pepsi." But many folks are still unaware just how versatile pork is on the table. The following few recipes show just what can be done with a bit of pork and a bit of human imagination.

Pork Cake

This recipe comes from our family friend, the late Mrs. Lora Momphard of Silex, Missouri.

Here sausage, black walnuts, and the wooden toothpick test combine to show its Midwestern roots. This cake is rich, very moist, and heavy, so you'll need some time after dinner to get your appetite up before sampling it.

½ pound pork sausage
1½ cups brown sugar, packed
1 cup boiling water
1 egg, beaten
2½ cups flour
1 teaspoon baking soda
1 4-ounce package lemon peel
1 4-ounce package orange peel
1 cup black walnuts
1 teaspoon cinnamon
1 teaspoon nutmeg
1 teaspoon cloves
1 teaspoon salt
1 cup white (golden) raisins

1. Preheat the oven to 350°F (180°C).
2. Mix together sausage, sugar, and boiling water. Let cool.
3. Add the beaten egg to the mixture.
4. In a separate bowl, mix together the flour, baking soda, lemon and orange peel, walnuts, cinnamon, nutmeg, cloves, salt, and raisins. Add to the sausage mixture, and blend thoroughly.
5. Line a baking pan with heavy brown paper. Add the batter to the pan.
6. Bake for approximately 1 hour and 20 minutes. The cake is done when a wooden toothpick inserted into it comes out clean. Let it sit in the pan for 7 minutes before removing it to a serving dish. Eat a slice of this on heavy china with a tall glass of milk or hot coffee, then go out for night chores.

Old-Time Pork Cake

Recipe courtesy of the late Mrs. Lora Momphard of Silex, Missouri

This is a local variant of a recipe you have to really get off the beaten path to acquire these days.

1 cup hot coffee
½ pound ground fresh pork fat
1 teaspoon baking soda stirred into 2 cups molasses
1 cup sugar
2 eggs
1 teaspoon allspice
1 teaspoon salt
6 cups flour
1 cup ground raisins

1. Preheat the oven to 325°F (160°C).
2. Pour the hot coffee over the pork fat.
3. Combine with the baking soda–molasses, sugar, eggs, allspice, salt, flour, and raisins, and mix well. Pour the mixture into a large Bundt pan, and bake for 2 hours.

Cracklin' Cookies
From the Lincoln County, Missouri, bicentennial cookbook

This is a rich, heavy cookie, with a unique texture.

3½–4 cups all-purpose flour
2 teaspoons baking powder
1 teaspoon baking soda
1 teaspoon cinnamon
½ teaspoon salt
2 cups ground cracklings
2 cups brown sugar
½ cup whole milk (cold)
2 large eggs, beaten
1 teaspoon vanilla
1 cup ground raisins

1. Preheat the oven to 350°F (180°C).
2. Combine the flour, baking powder, baking soda, cinnamon, and salt in a large bowl.
3. Cream the cracklings and sugar in another large bowl. Add the milk and stir until thoroughly combined. Add the eggs and vanilla, and stir to incorporate, then add the flour mixture, and stir until combined. Stir in the raisins.
4. Shape into balls. Flatten out, and bake for 10 to 15 minutes.

Cracklin' Corn Bread
From the Lincoln County, Missouri, bicentennial cookbook

This is a most hearty form of corn bread that holds body and soul together.

1 scant cup cornmeal
Pinch salt
½ cup cracklings
Boiling water

1. Preheat the oven to 400°F (200°C).
2. Grease a 9-by-9-inch baking pan.
3. Stir the cornmeal, salt, and cracklings together, and add enough boiling water to make for easy spreading.
4. Spread in the prepared pan, and bake for about 45 minutes.

Childhood Piecrust
The roots of this recipe are lost to time, but it is the secret behind more state fair–winning pies than you can shake a stick at. A favorite boyhood snack of mine was simple strips of piecrust dough (the scraps from the pie baking) sprinkled with sugar and baked in the oven at the end of baking day. This recipe will offer up a few scraps for those extra treats.

3 cups flour
1 teaspoon salt
1 cup lard
½ cup water

Combine the flour, salt, lard, and water in a large bowl, and mix together thoroughly. Roll out as you would any other piecrust.

5

FITTING AND SHOWING HOGS

Bigger than a lamb but much smaller than a club calf, the hog is one of the most popular of all animals for youth-project work. In fact, youth-project work is probably one of the most common uses for hogs (after home meat production) on small country places.

Hog production is quite strongly identified with the Midwest and the Corn Belt, but many of the major American swine breeds, such as the Chester White and the Duroc, were developed in the eastern United States. Hog shows in Oklahoma and Texas often draw entries of a thousand head or more; in California, too, hogs are a popular project for young people who have very small plots of land from which to work.

A project pig, ready for the show-ring, will weigh between 220 and 260 pounds — and yet it is probably easier to handle in the ring than either a calf or a lamb. Unlike its cartoon image, as I've said before, a hog is neither a dirty animal nor a glutton. First-year 4-H kids as young as 8 regularly show one or two animals in market-hog classes in livestock shows from Maryland to Oregon.

Showing Market Hogs

In youth swine work, there are basically two types of projects. One is the market-hog project, which is the feeding to market weight of 1 to 10 head of feeder pigs. The other is the gilt-and-litter or sow-and-litter project, in which the young person takes the female from mating to the birth of her litter and then takes her pigs from birth to market weight or sale as feeder pigs. Both are good learning experiences, though the former is the normal choice for a beginning youth project.

During my Future Farmers of America (FFA) years, our home-county fair had a

FITTING HOGS

In its broadest sense, the term "fitting" is used to describe the entire process of readying an animal for exhibition. A hog is "hand fitted" for an appearance in the show-ring. Used in its narrower sense, **fitting** means those final steps taken to get the hog ready just prior to entering the show-ring. A washed and brushed hog is a "fitted" hog.

Show day. Youngsters are dressed for the show-ring, and a Duroc hog is fitted and awaiting his turn before the show judge. The girl is gently rubbing the pig with a show stick to keep him calm.

ADVANCING TO GILT- AND-LITTER PROJECTS

By their third or fourth year in 4-H work, and second or third year in FFA work, I like to see youngsters move up to gilt-and-litter projects to gain further learning experiences and skills. Granted, the two projects seem to have switched positions in importance over the last few years, and a greater number of young people now participate in the simpler market-hog projects.

daylong breeding-stock show for pigs farrowed that spring. Entries would often number several hundred, and some youngsters showed the production from as many as 8 to 10 purebred litters. The market-hog class came quite late in the day and generally drew modest numbers. Now breeding-stock projects are far fewer in number, and market-hog projects have grown more popular. At some county fairs, three hundred or more hogs may be entered in market shows and rate-of-gain competitions. At our local fair, the market-hog show may take six to seven hours to conclude.

Since they tend to be more popular today, market-hog projects will largely be the focus of this chapter. Information on showing breeding stock, however, appears later in this chapter, and details about raising breeding stock appears in chapters 7 through 9 for anyone interested in litter projects.

Market-Hog Projects

A market hog is a barrow or gilt that enters the show-ring in the weight range typical of hogs sent to market: 220 to 260 pounds. The typical 4-H or FFA market-hog project entails feeding out from one to three pigs with a specific county fair or market-hog show in mind. A market-hog project has modest space and financial requirements. Good show-ring candidates are probably easier to find among hogs than any other major livestock species—which is a testament to the quality bred into today's meat hogs.

Few days are busier, more hectic, or more exciting in a young person's life than show day at the county fair. Months of work and preparation are about to be put to the test, including the young person's selection skill, ability to care for the animal and dedication to that task, and knack for exhibiting.

SHOW ME WHEN?

Although there are still some spring barrow shows and some of the national competitions are scheduled in the winter months, most market-hog shows fall in the fair season of July through September, and most state fairs and exhibitions happen from late August through October. The pigs for such shows are born from December through the first four months of the year.

In the South, East, and Southwest, shows tend to run later in the year; September and October are the primary exhibition months. The West Coast show season more closely approximates that of the Midwest, with a bit of a later start and an earlier finish. Things wrap up in the Midwest in November with the American Royal at Kansas City, Missouri.

The most common practice in a market-hog project is to feed out a hog or hogs for participation in a specific show or event. At the 2016 Lincoln County (Missouri) Fair, nearly 150 youngsters prepped more than 350 hogs for a local county fair show. How many of those head made it to the show-ring depended on such variables as their genetic potential to grow quickly; their physiology and ability to remain sound; how the weather affected their comfort, metabolism, and appetite; how their health and nutrition were managed; and how much time and consideration were given to their care and well-being.

Show time. This young lady has her hog properly positioned for best visibility.

Project Goals

The purpose of market-hog youth projects is for the children involved to learn such skills as good animal selection and stock feeding; to gain knowledge of livestock marketing processes; to share in positive group activities with other young people; and to get a good taste of life in the real, very competitive world in which we all now live. I showed hogs and bred purebreds during my own FFA years, have been involved in the purebred swine sector for more than 30 years, and led the market-hog project for our local 4-H club for several years.

At shows, the animals should enter the ring as good representatives of what market hogs are meant to be. They should be the correct weight, the correct age for their weight, and of good overall type with the correct amount of finish. The market hog is the primary product of the pork industry, and because of that there is always a great deal of interest in the market hog classes at any livestock show — in fact, they often draw some of the largest crowds of any fair event. Our local market-hog show is actually broadcast live over one of the area radio stations.

The shoat in a market-hog project has a twofold task: to eat and to grow. It is the young person's responsibility to see that the shoat is kept comfortable, safe, and healthy while fulfilling its tasks.

The project begins with a rather young animal, just 8 to 10 weeks of age. As the animal grows and matures, its feed and care needs will certainly change. It will be on feed for only 100 to 120 days, one of the shortest terms of all youth livestock projects.

In the end, though, the real goal of any youth livestock project is to produce "blue ribbon" youngsters. As an old-line vocational

SHOWING PIGS VERSUS CALVES OR LAMBS

A club calf will cost far more to buy than a project pig, which sells for between $1 and $8 per pound. A club calf will also be on feed for far longer (and consume much more feed), will be more difficult to fit and train for the show-ring, and will be more difficult to house and transport. Its sheer size can be imposing to many youngsters.

Growing out a club lamb takes a little less time and costs less in feed than raising a pig. Still, the costs to buy are comparable, and in many areas high-quality lambs are not nearly as available as showable pigs. The fitting needs and equipment requirements are greater for lambs than for pigs. With a market hog, there is no need for a fitting table, shears, or combs. The basic hog-show kit now includes little more than a garden watering can and a stiff-bristled scrubbing brush.

The un-fun side of the show: keeping the pens and animals clean while the show is running and then completely cleaning the barn at show's end

Hogs being readied for the show-ring are pushed for optimal growth in what can be some of the hottest weather of the year. Preferably, locate their housing in an area that is well shaded by mature trees with prevailing breezes flowing through.

agriculture teacher once told me, perhaps the best ribbon of all for a beginning youngster to receive is a white one. If this doesn't fill a child with the resolve to work harder and do better next time, then it will serve as the incentive to move on to something else, where youthful talents will be more focused and better applied.

Have Hogs Show Ready

The time spent in the show-ring is very short when compared to the time spent selecting, growing, and fitting a market hog for its moment on the fairground's **tanbark**. On show day, it should be market ready in every way.

Housing and Equipment

Housing for hogs being readied for show need be neither expensive nor elaborate. I can recall seeing more than one $3,000 gilt drinking from a $3 rubber pan and snuffling through the dirt of an open lot for stray grains of corn like any other hog.

The facilities for feeding out a small number of hogs outlined in chapter 3 should be more than adequate for holding and readying two or three head for show. Show hogs will be pushed for optimum growth in what can be some of the hottest weather of the year, however, which is an important consideration.

Toward that end, I like to see housing for show pigs placed in an area that is well shaded by mature trees and where prevailing breezes can flow through. You must have airflow to cool growing hogs in hot and humid weather; to enhance it, use a house with doors on both ends or a back panel that can be raised and lowered. These can be opened in proper sequence to facilitate airflow through the house and across the hogs.

Likewise, roofs that lift or slide back can be positioned to bring more airflow across the hogs. And certainly don't pack the hogs into the housing in warm weather. Allow them something on the order of 12 to 16 square feet each of indoor sleeping space, plus that much or more on the outside pen floor. Drylots are even better for maintaining soundness and muscle tone in pigs being readied for show; allow at least 150 square feet of space apiece. To further develop muscle tone, be sure to place feeding stations, watering equipment, and sleeping areas well apart from one another.

Once they are penned together, try to avoid moving the pigs to different enclosures, breaking up the original group, or introducing new pigs to that group; any of these can cause stress and injuries from fighting. If the weather should turn very warm, put a sprinkling hose on a timer and position it above the hogs. Set the timer to turn on the hose for 5-minute intervals to mist the hogs lightly. Keep this up during the hottest part of the day, generally from 10 A.M. to 4 or 5 P.M.

Feeding for Show

Feed is the fuel that propels a hog's growth. If you are readying a hog for show-ring competition, you need high-octane performance.

Many producers achieve this by holding show pigs on a 15 to 16 percent crude-protein ration throughout the entire feeding period; I have even seen pigs fed for show on 17 to 18 percent crude-protein rations. Even the 3 percent increase in performance that comes from feeding higher-priced, pelleted complete rations is justifiable when readying a hog or hogs for competition, especially if you are going to be participating in a rate-of-gain competition.

Feeding challenges. Going into hot weather, palatability and consumption are two very real problems to be resolved. In very hot weather (temperatures in the mid-80s and higher), the simple physiological act of metabolizing food into energy the hog uses to function adds to a hog's heat stress and thus decreases his appetite. Because of this, rations in hot weather should be made as palatable and nutrient dense as possible.

The easiest way to do this is to add fat to the ration. For example, some producers mix in feed-grade fat at the rate of 3 to 5 percent of the total ration. Rations can be top dressed with an inexpensive liquid cooking oil or even the discarded cooking oils or greases from a local restaurant. Just be sure to start the hogs on fat slowly and build up its level gradually. A too-sudden shift in ration content can cause gastric upsets, with quite serious consequences for pigs on the fast track to a livestock show.

Pelleting feed also seems to add something in the way of extra palatability; the heat and pressure applied to the feedstuffs during the pelleting process should improve their digestion and utilization by the animals as well. Increasing the ration's protein content may also improve growth rates for animals that appear to be bogging down as temperatures increase. Try hand feeding to appetite if just one or two pigs are being readied for show. An old trick is to gradually build up the growing ration with milk-rich pig starter until the animal is consuming a pound or so of the **creep feed** offered to nursing pigs mixed into its ration each day. It's a costly practice, but it's often the best way to achieve much-needed rapid weight gain.

Some producers encounter a problem that is the exact opposite of slow growth: They have a pig growing at a rate that will put it past its prime weight by show time. They, too, will benefit from some hands-on dietary management.

Begin by pulling that self-feeder and going to hand feeding. Feeding to 90 percent of appetite will help to check growth; some folks pare daily consumption even further. You might also begin by shaving protein levels rather than amounts fed. However, note that in the latter stages of the finishing period you must be wary of either throwing the hog into a complete stall or slowing its growth curve and causing it to pack on more fat rather than lean gain.

Be sure to make any change in the ration a gradual one, and couple any ration trimming with a course of exercise for the hog in question. Walk the animal in and out of the pen for 30 minutes or so in the cool of the day to condition it, tighten its muscle tone, and get it used to being driven in the show-ring.

SELF-FEEDERS

A self-feeder should be your feeding tool of choice, because it will keep feedstuffs before the growing hogs constantly and also maintain the feedstuffs in a cleaner fashion. A small wall- or gate-mounted, one- or two-hole feeder is probably your best bet, because it facilitates frequent ration changes. You are pushing these growing hogs rather hard for a specific event and so should not skimp on either equipment or feedstuffs.

Selecting Hogs for Show

While a rose may always be a rose, pigs are born quite different from each other. Selecting a pig or pigs for show is not as simple as buying the first thing that walks by with four legs and a corkscrew tail. In fact, there is much more to take into consideration than when you are selecting a hog for market or for home processing. Here you are truly trying to skim the cream off the local pig crop.

The best advice I can give on show-pig selection is to go out and take in as many nearby hog shows as possible. Most states have early spring–type events and judging days for 4-H and FFA youths. Try to catch your local hog-show judge in action, to learn his or her style and selection criteria. At the shows, note carefully

DON'T FORGET THE WATER!

Do not neglect the most important of feedstuffs: drinking water. In warm weather, it is vital that drinking water be kept clean, fresh, and appealing, to encourage maximum consumption. Keep it available to the hog in the cooler parts of the day when hogs are fed, which will do much to encourage consumption at proper levels.

LIVESTOCK SHOW HISTORY

The livestock show's roots reach all the way back to ancient times, when early nomadic herdsmen would meet to exchange breeding animals and the young of the year to add new blood to their herds and flocks. This tradition continued through medieval market days and beyond. Early in U.S. history, there were displays of various livestock breeds and varieties at harvest fairs and farmers' gatherings.

The competitive aspects of livestock showing came later. Over time, the side-by-side comparison in the show-ring came into its own as a primary arbiter of desirable livestock type both in this country and abroad.

the judge's comments as he or she explains the class placings, and note carefully the animals selected as class and overall winners. A judge doing a good job will talk freely about what he or she is looking for, and the consistency of the judge's opinion should hold up through all the placings. The 240-pound class winner should look like a slightly older, slightly larger version of the 220-pound class winner.

You can also go on the Internet to find pictures of breed association standards and top winners for the year before. Compare the looks and type of the breed standards and big winners to any sow you are considering, but be aware that type trends can change rapidly.

Desirable Hog Traits for Showing

A pig for showing has to be very typey. This is a vague term, to be sure, but what it most specifically means is that the animal demonstrates large amounts of muscling, is very lean and trim in appearance, is of exceptional length, and is otherwise strong in the current trends in swine type. The trends are dictated primarily by the breed registries, the livestock shows, and the industry demands.

That latter point, the seemingly ever-varying trends in swine type, is the true kicker in the showing of hogs in any class or category. In the twentieth century alone there were close to 20 major changes in swine type, each one believed to have moved the animals closer to an ideal in meat, reproductive, or growth type. Among the functions of the show-ring are to showcase such changes and then see how well they hold up over time and in head-to-head competition with other hogs. It thus becomes imperative that those selecting animals for the show-ring be well read on the subject of swine type, attend a number of shows, and mine their project leaders for as much hard information on swine type and show-ring conformation as possible. Hit the swine press (the breed association magazines, *Showbox* and *Purple Circle*) hard for information on emerging trends in type and show-ring developments.

Sizing Up the Judges

Normally, market-hog judges come from one of four groups: purebred-swine producers, university staff members or Extension Service personnel, and representatives of the meat-packing industry, such as order buyers. There is much to be said for knowing a particular judge's likes or dislikes during the selection process. One of the best examples of this is how the soundness of the animal and its effect on mobility are addressed.

TIPS FOR SELECTING SHOW HOGS

There is a body of nuts-and-bolts data that can serve you well when you're selecting hogs for a market-hog competition. Following is a sample:

- Gilts generally grow a bit more slowly than barrows but nearly always hang a leaner carcass on the rail. Their leaner nature does make it possible to push them harder with a ration higher in crude protein.

- Select the largest and most well-developed gilts in a pen; they demonstrate a higher growth potential than their siblings.

- In some areas, market-hog competitions are still called barrow shows; castrates are certainly the bedrock on which the modern pork industry is built. To show a barrow, select again from the largest pigs in the pen, but also for a trim appearance about the head and along the underline. The presence of at least three nipples ahead of the penile sheath is a good visual indication of an animal's potential to hang a long carcass.

- Crossbred pigs tend to grow faster than purebreds, but to any rule of thumb there are exceptions. Right now, stylish purebreds showing a lot of breed character are a rare enough sight in the show-ring to catch a judge's eye and earn a long second look as the classes are placed.

- When selecting crossbred animals, load up on the colored breeding (black and red) as much as possible, to capitalize on the muscling and growth characteristics for which those breeds are noted.

- Consider the lighting that will be used at the event you plan to attend and the time of day the show will occur. Black and red hogs appear to show up better under artificial lighting. White or mostly white hogs seem to show better in daylight.

- If possible, when you enter the fairgrounds, weigh in at least one backup pig for each animal that you plan to show.

Some judges are a little more forgiving of things like uneven toe points or slight defects in leg structure than are others. Don't take such things as givens, however: These are often the same features other judges use to resolve final placings. There's little of economic value in a front leg or a hoof point, but an animal that doesn't move freely and won't take its place easily at the feeder consequently won't grow as efficiently as it should.

At many shows, judges sign multiple-year contracts; you can thus develop a real feel for two- and even three-year judging trends at such shows. Many states also sponsor early-spring symposiums on livestock judging and evaluation on their agricultural school campuses. These can be good guides to current trends in popular and winning livestock type. Contact your local Extension agent to learn if such events are held in your state.

WATCH, LISTEN, AND LEARN

At the local fair, the place for interested youngsters to be when they are not actually showing is in the ringside seats, listening and watching as the judge places the other classes in the show. There are open shows for market hogs in which youths often compete, but most shows are for breeding animals. These are extremely competitive, are dominated by veteran breeders and showmen, and have elements of combat. It's probably best if young people sit these out, learning what to do and what not to do at a competition.

This judge is evaluating hogs and young showmen alike and may ask questions of the youngsters as he moves through the class and assesses the hogs.

Leading Show Breeds

At this time, Hampshire-cross pigs stand head and shoulders above anything else in the field of market-hog competition. The strongest crosses are the Hampshire × Duroc F_1, the ¾ Hampshire/¼ Duroc cross, and the Hampshire × Yorkshire F_1. Berkshires and Landraces are also being used in breeding to formulate some snappy genetic combinations for the show-ring. A **growthy** gilt with a lot of length and internal dimension and wrapped in

a good black hide is going to be hard to beat in market-hog competitions up to about the 260-pound class.

Where classes are carried to extremes, up to 300 pounds, the edge seems to tip back to barrows and to hogs that carry more white breeding, for the trimness they can maintain at such weights. Start with as heavy a pig as you can in case there is a problem with warm weather and growth slowdown.

PREPPING FOR THE SHOW

About a month before the date, your preparation should really kick into high gear.

Get the hog used to being handled! In the cool of the morning or late evening, walk the hog on different surfaces around the pen and farmyard to get it accustomed to being exhibited. Acquaint the animal with both the show stick and the brush. Use light taps on the shoulders with the show stick to keep the hog in motion and turning properly at the corners or ends of the ring. If you tap the pig on the hams or around the loin, it will unwind, or go into a sort of clinch, then sag or hunch up with its tail uncurled, which is not a pretty picture for the judge.

Get the hog comfortable with being brushed. A brush is the tool for cleaning the haircoat; in the ring brush away bits of dirt, sawdust, or straw from the hog. Older youngsters can often show a market hog with nothing but a brush in the show-ring. I don't like to see youngsters older than 13 go into a ring using a show stick, hurdle, or cane, because this suggests they haven't spent enough time getting their animals ready. Further, striking, kneeing, or kicking an animal is sharply discounted by nearly all judges.

Begin evaluating the hog's emerging growth and finish curve. Do gains need to be pushed, or does the feed need to be scaled back to produce a trimmer-appearing finish? Twice-a-day walking may be necessary to increase trim or tone, and a ration switch as extreme as going down to just a few pounds of shelled corn daily may be in order. To push gains, bring on the pig starter or switch to slop feeding — the mixing of 1 part complete ration to 1½ parts water. This should increase warm-weather consumption, add to body fill, and increase finish.

Take an honest look at the hog, and seek an outside opinion as to its merit. Ask a project leader, a teacher, Extension agent, or local swine producer to help determine show-ring strategy and which of the animal's better traits to play up in the ring. A couple of years ago, one of the youngsters in our Silex Flyers 4-H club had a very good Hamp-cross gilt with a bad tendency to bunch up when she stood still. To show well, she had to be kept moving in the ring. She was selected reserve grand champion, but had the child not let her get jammed in a corner of the show-ring and bunch up in front of the judge, she would have been grand champion. An outsider can help you recognize such problems, along with the means to correct or counter them.

Get cracking on the paperwork. Get in your entry forms, and make an appointment with the vet to have blood tests taken so you'll have those health papers in hand before the last-minute rush. You're likely to need written proof that your animal is free from pseudorabies and brucellosis.

Going to the Show

At most shows, you are expected to have the hog or hogs on the show grounds between 12 and 36 hours ahead of showtime. They are generally weighed in as they come off the truck or trailer. That weight is then used to determine which class they will be shown in, along with their placement in the rate-of-gain competition if it is a part of the show. This is also the time when ultrascan or "Real-Time" readings may be taken on the live animals, to measure things like loineye size and backfat thickness, which are further aids in placing the swine classes.

With fairly heavy hogs, it is a rather widespread practice to withhold both feed and water for a period of 6 to 12 hours prior to morning weigh-in, generally through the night. It puts the hog across the scale with an empty bowel and bladder, and a couple of pounds of body shrink, which may allow for a go at the lighter weight classes, where success means more. The animal will regain most of the lost weight as soon as it goes back on feed and water. I neither condemn nor condone this practice; it is very widespread, there are no specific show rules against it at most events, and it seems to have no harmful effect on the animals (feed and water are often withheld from hogs at farrowing for this same length of time).

I do caution against withholding drinking water during very warm weather. And at the opposite pole, keeping lightweight hogs on feed and water right up to when they go on the truck may be necessary to make show weight.

Transport

To spare the animal any stress or risk of injury, it needs to have a soft, sweet ride all the way to the fair or show. Do the trucking early in the day, while the coolness of the night still lingers. If it is very hot, a 4-inch layer of damp sand or sawdust in the truck or on the trailer floor will help keep the hog more comfortable. A note here: I once exhibited a pair of Duroc gilts that developed hives — an allergic reaction to the oak sawdust bedding used at a local fair. Experiment with any expected bedding changes at home first.

Move the show animal in a very slow and easy fashion. Use no whips or prods, no kicking or kneeing, and move the hog by blocking behind it with a hurdle or short gate. Entice it with feed, and let it move along at its own pace; a running hog is a hog in peril of falls and other injuries. And unload it in exactly the same way. That tail is not a handle for lifting and twisting.

Show-Box Supplies

As an exhibitor, you will also need to take along a show box packed with the essentials for a day or a week at the show. This is generally a footlocker- or trunk-type affair made from heavy plywood and painted to look attractive in the back of a pen or along the alleyway in the exhibition hall or show barn. White, blue, red, or green — traditional colors for farm buildings and equipment — is often the color of choice, and many folks further adorn their boxes with decals or stickers of favored breeds and the name of the family farm in large letters. I've seen show boxes large enough that small youngsters can actually sleep atop them.

Into the show box should go:

- Nested feed and water pans
- A sack or two of feed
- A garden hose for watering and washing hogs
- Brushes for cleaning the animals
- Dishwashing liquid or other liquid soap to wash hogs (dishwashing liquid will also rid hogs of lice without causing any residue or withdrawal problems if a louse problem should emerge at or close to showtime)
- Folding cot or sleeping bag, if you are required to stay with your animals overnight
- Show stick
- A small hurdle or two
- A pen sign or card to identify yourself

Grooming Hogs for the Ring

At one time, swine exhibitors applied colored oils to their colored hogs, to bring up the haircoat and to make it glisten in the show-ring. White hogs were liberally sprinkled with a white talcum-type powder for the same reasons. Such practices have since been eliminated, to promote livestock shows and exhibitions as showcases for the animals' merits and not the fitters' skills and tricks.

I recall once helping a fellow FFA member show his Chester White hogs. On the way to

Market hogs being readied for show should become accustomed to brushing before showtime.

the show-ring, the white gilt I was handling was accidentally splashed — liberally — with muddy water from a puddle between the barn and the ring. The powder on that gilt began to set up like concrete. I drove her into a corner of the show-ring, knelt down, and began to brush her haircoat frantically. The harder I brushed, the more the powder set up. Just then the show judge gave me a bit of a nudge with the toe of his boot. He pointed to the center of the ring and said, "Young man, we're having the hog show over there." As it turned out, I got into the show and my "dappled" Chester White won a red ribbon.

Soap, water, and elbow grease are just about all that are now applied to a hog to prepare it for a show-ring appearance. Scrub the pig downward from the centerline of its back. A stiff-bristled laundry brush will do a good job of freeing dirt and manure from a hog's coat. Once you're back in the pen and awaiting the call to the show-ring, a garden watering can will hold enough water to keep your brushes damp for last-minute touch-ups.

At most shows, the sprinkler can even be carried to ringside and used to drizzle a little water on the hog's snout as a refreshing measure. Also useful in the ring is a small scrub brush that can be carried in a hip pocket or free hand. Water can be useful in keeping a hog cool — but never apply water to a hog showing any signs of heat stress. Cold water heightens stress in an animal that can't sweat and must pant to shed heat. Cool the hog down gradually with slow use of room-temperature water on the face; otherwise shock and death may result.

Showtime

The long months of selection, preparation, and fitting climax in the show-ring, and the action

ATTIRE FOR EXHIBITORS

The exhibitor also needs to prepare for the trip into the show-ring. A judge I very much respect makes a point of announcing at the start of every show he judges that he will be reading caps and T-shirts. If he doesn't like what they say, he adds, it affects his placements.

In the show-ring, a shirt or blouse of simple pattern and Western cut along with dark jeans or slacks are the best choices for attire. Young people with long hair often tie it back so that it won't be bothersome in the ring. Shoes or boots should be comfortable, support feet and ankles well, and be sturdy enough to provide protection from all those hooves.

A young typey Hampshire gilt and a young man proudly wearing his FFA shirt

there, can be rapid and confusing. You need to go into the show-ring with a well-thought-out plan to show your hog to its best advantage.

- Know the pig's strengths and weaknesses, and be prepared to put the pig in its best light. Subtly use the show stick to guide the judge's eye to the strengths.
- Do not be the first exhibitor into the show-ring, or the last. Try to position yourself to be among the first half of the exhibitors into the ring. Enter the ring with your hog solidly under control and moving in the same pattern as the others there.
- Never come between your hog and the judge.
- The judge seeks views of the hog from head-on, from both sides, and from the rear. To get that head-on view, the judge will be in position to see each hog as it enters the show-ring. Some judges

even request that the hogs be brought into the ring one at a time rather than in cavalry-charge fashion through a wide-open gate.

- As much as possible, keep the hog's strongest features before the judge. If you have a hog with good hams, set it up for the judge to get a good view of that economically important trait.
- Keep the hog moving, and don't get caught in traffic jams in the ring ends or corners. A lot of judges watch very closely as hogs turn, because that is where problems may be most apparent.
- Be prepared to answer questions as to your hog's age, its breeding, and the rations fed to it. These questions may be used to break ties and to determine showmanship placings.
- Follow the judge's directions in the ring closely.

In many shows, the judge will have the top hog picked within five minutes of getting all the animals in that class into the ring. The good ones show up rather quickly in most market-hog classes. The difficult task is sorting through the remainder after the top two or three have been pulled from the group.

Deciding where to break between blue and red ribbons and red and white ribbons can also be difficult for a judge. Ranking red-ribbon winners, when required, can be very time consuming.

Top pigs are generally directed to small pens alongside the show-ring, and red- and

AS THE JUDGE SEES IT

In my junior year in high school, I served on our FFA livestock-judging team, and for many months we were drilled on what to look for when evaluating breeder stock, feeder stock, and butcher animals. We honed our stockperson's eye by formulating mental images of what the good ones should look like and then comparing the animals before us with those mental images.

We were told to never make negative statements but to give positive reasons as to why we preferred one animal over another. The animals before us all belonged to somebody, and that person might find negative comments hurtful, and after all, our rankings were really just our opinions.

Fault finding is part of livestock evaluation, but it is often the most difficult. The best judges will tell you that they have the top animals in a show class within minutes, but when they have to rank every animal in a class they bog down. It's tough to differentiate between the 23rd and 24th pigs in a class of 25 head.

The place to begin is with **culling** out the animals with the most obvious defects. These include animals with poor or incorrect breed type that have obvious health problems, that are unsound, or that are too extreme in muscling or leanness. The ideal is an animal that is middle-of-the-road for all traits or one that is exceptionally strong in one or two traits where the existing herd needs upgrading, while being of good type in other areas as well.

Size for age is a good indicator of vigor and will to thrive. Litter size at birth and weaning tell much about milk and mothering, and performance data, if obtained in a realistic way, will help to guide selection. Short steps, a tucked stance, squealing and other signs of pain or stress, a lethargic manner, slashing and biting, and an asymmetrical body appearance can all be clues to a lesser animal. You can almost sense them before actually picking out points of fault. The simple truth is that the stockperson's eye has to be honed, and that means eyeing up as many hogs as possible in as many places as possible.

I was raised as a bit of a sale-barn rat and thus saw thousands of head sell at auction. I attended numerous breeding auctions back in the day, participated in numerous live and picture-judging contests, and went to more than my fair share of hog shows.

My running buddies and I had a game we would play at breeder auctions. There, animals often would be offered in large groups, with first choice going to the highest bidder and continuing downward in order of bid until all were sold. The game was to see who could pick out the choice animals before the winning bidder called out his or her picks.

white-ribbon winners sent back to their respective pens in the barn. On the way out of the ring, each hog's paperwork will be turned over to the show secretary, and ribbons will be dispensed.

As long as hogs remain in the ring, though, the show is still on, and exhibitors need to keep their hogs moving in proper fashion. As I've noted, a good judge will give reasons for his or her placements and may even take longer with some hogs lower in the order than with class winners. Some very extreme hogs can present special problems in their placement, and a good judge will address this in his or her remarks.

It was fast competition! You had to learn to see an animal quickly as the total of his or her parts. And such a competition becomes even more exacting when you are the winning bidder.

I wish there was a fancier way to say it, but you simply have to learn to see them, really see them. Form that mental picture of a good one, and then hold onto it when the bids are flying and 50 head are running past you, or you and your hog are standing under the hard eye of a professional judge. After all, the longest gilt in a class of 25 Duroc gilts is a lot easier to spot than the shortest gilt among the three shortest gilts in that grouping.

This judge is getting down on hog level for a better view of bone structure and soundness. Both traits are also quite important to market hogs.

Show Schedules

Most market hogs show in classes based on their liveweight, with a weight range of no more than 5 to 10 pounds among all the hogs in a given class. Class first- and second-place winners will be called back later in the show to compete for first and second places in their division. (Generally, a division encompasses the hogs from three to five weight classes.) At most shows, there are two to four divisions.

From the division winners are selected the grand champion and reserve grand champion of the show. In divisions and finals, the overall winner is chosen first. After that, the animal that stood second to it in its class or division is brought into the ring to compete with other first-place winners for the second-place position. Staying in the hunt until the last animal is placed can make for a very long day or evening. Our county fair show has often started at 7 P.M. and continued until 1 or 2 in the morning.

With the completion of show-ring competition, there may be another day or two of show-related activities to work through before a fair or other event is over. There may be a requirement to keep the barns full for later fair visitors, a sale of show winners, or even an auction of all show participants. At many shows, all the animals and gear must be removed, the pens cleaned, and the barn set back as it was before premiums and sale checks are paid to the participants.

Until the show is over and the animals are clear of the barn, they have to be the young exhibitor's first priority. No matter how much the Ferris wheel and cotton candy may beckon, pens and alleys need to be kept clean, hogs monitored to ensure their safety and comfort, and passersby made welcome. Be sure to answer any questions from those visitors; many are potential bidders in the auctions that often follow hog shows.

KEEPING YOUR PERSPECTIVE

Sitting astraddle a bale of straw in the back of the show barn, you can learn many lessons: how to feed a winner, where the best pigs are to be found, how to do a better job of fitting for the show-ring, and so much more. And one of these lessons should be kept in mind for as long as you pursue a career of any sort out on the tanbark: The judge's placing is just one person's opinion.

I have seen hogs with a blue ribbon in one competition, which then do no better than a red ribbon the following week under a different judge. A properly run show is first and foremost a learning experience; it is the latest in swine type presented for honest evaluation before the eyes of other swine producers. The ribbons and premiums are but a pleasant by-product — or are supposed to be.

To me, the hallmark of a good hog show is that moment when, on the way out of the ring with a lower-finishing hog, a youngster turns back for a moment to shake hands with the exhibitor of that day's winner. On another day, with another judge, the placements might very well be reversed. But the friendships made and the careers launched inside the show-ring can last a lifetime.

Showing Breeding Stock

Many young people can and do go on to breed and show their own hogs and even to compete in shows for crossbred and purebred breeding animals. Breeding-stock competitions can include county fairs, state fairs, and conference-type competitions at which breeders from across the nation come together to compete.

The basics of breeding-stock competitions are the same as for market-hog shows, with one exception: Both the handlers and the animals in these competitions tend to be older. Hogs a year or more of age are still invited to compete at some state fairs, a few of which still conduct big-boar contests. These include boars 5 years old and older, who often top one thousand pounds. "Open classes" at these fairs are open to all comers. Nearly every year at the Missouri State Fair, for example, one breeder or another is honored for 50 years of competition at that grand and classic swine show.

While many of the participants at the state-fair level are veterans of decades of such competition, most state fairs also maintain breeding-stock and market-hog classes just for 4-H and FFA youngsters. These young people are also generally free to go on and compete in the open classes. By the time the state fairs roll around each year, most of the animals of the year going into the largest breeding stock classes are 8 to 10 months of age.

Breeding-Stock Classes

Breeding-stock classes are most often divided by animal age, in roughly 30-day intervals. For example, an early-spring Duroc class might include just boars farrowed between January 15 and February 15 of that year. In open shows there will be breeders who farrow

THE STAKES ARE HIGHER WITH BREEDING STOCK

State fairs and national conferences are true arbiters of swine type, and at some a lot of dollars are on the line. A big win can lead to a selling price in the healthy four- or five-figure range for the animal in question. Careers and breeding lines have been launched on just one or two such wins, and these shows are very much merchandising devices for bloodlines and breeders.

many litters per year and thus can exactly match animals to each class. Returning to the above example, a boar farrowed on January 15 will be a full month older than the youngest members of its class. Given the rapid growth rate of modern hogs, this can provide an animal with a tremendous advantage in the visual appraisal that is still so much a part of breeding-stock judging.

Boars that are about 180 days old are moving into sexual maturity, and one of the most common of the secondary sexual characteristics in boars is aggressiveness. Between boars raised as penmates, this is no problem, but boars of this age driven into the tension of a show-ring and confronted with other, strange boars can and often do mix it up. A boar fight can be a simple inconvenience or a costly tragedy. More than one young boar has had to be destroyed following a show-ring mishap.

The Value of Stock Shows

One of the tests I use to determine the sincerity and dedication of young would-be hog raisers and showmen is to take them to a nearby FFA breeding-stock show and sale. If, after seeing one or two teenage boys bounced around by some scrappy 350- to 400-pound boars, they retain their interest in hogs, they may have what it takes to take their own licks in and out of the ring. Actually, I have seen lambs and calves take a harder toll on youngsters than hogs. Still, this is a problem every child will encounter sooner or later.

Not nearly as many seedstock producers take the show-ring route now as in the past, as markets are often based on genetics and testing data, but it is still an important venue for swine promotion. The big swine shows are followed and reported on as much as many sporting events, and the results are as eagerly awaited. Winners at an event such as the World Pork Expo or the National Barrow Show can define the industry for years to come. Within a relatively few months, their sons and daughters can be going into use in herds all across the country.

I'm a true fan of hog shows and especially enjoy seeing breeding stock being shown; there is no better place to practice and refine that all-important stockperson's eye. These shows put to a hard test the mental image of a good hog that we all carry around in the backs of our minds. Little does more to raise my estimation of a judge than hearing him agree with my opinions. And little does more to make me a better producer than a judge's comments that cause me to really think about what it takes to make a hog a good one.

There are controversial aspects to the show-ring, however. For instance, some folks question just how meaningful eyeball comparisons of hogs really are to modern commercial producers. The judging at shows is largely based on eye appeal and does not always factor in things like real carcass data or reproduction-performance traits. Others wonder if show-ring events are too competitive and too prize oriented for youths.

It's true that in many instances show-ring competition is not set up to evaluate such important aspects of swine production as reproductive performance, exact growth rate, feed efficiency, and various carcass traits. And exhibitors with all different levels of experience compete in most local swine classes. It is not uncommon to see an 8-year-old, first-year 4-Her in the ring competing with young people in their late teens with 10 years of show experience. Still, most of those teenagers began their show-ring careers in the same way as the newcomers, and all are operating under the watchful eyes of show officials.

It is also possible to "buy" a blue ribbon: Some families have more money to invest in project animals than do others.

Finally, there are, unfortunately, some show-ring cheats. These are exhibitors who attempt to show overaged animals; abuse certain animal-health products, such as steroids; or employ the skills of professional fitters and showmen to ready their animals for the show-ring. Most abuses can be spotted by show officials, though, and more and more show animals are being forced to submit to health examinations to detect and counter such abuses.

JUDGED NOT BY A COMPUTER BUT BY THE PUBLIC

In this age of extensive production testing, ever-changing scanning devices, and the use of computers to project estimated breeding values and progeny differences in seedstock, many folks question whether the show-ring retains real value in livestock selection any longer. I share a belief with a great many others that, even if it is among the oldest selection and comparison tools available to us, it is still a good way for livestock to be evaluated and one very much in keeping with the character of the American family farmer.

The show-ring provides us with that all-important and objective outside opinion; allows our opinions to be confirmed or disproved by head-to-head competition before the general public; and, in all but rare instances, its animals truly rise or fall on their own merits. In the modern livestock game, it has been proven to build character, legends, and herds and flocks that shaped the trade for generations.

Displaying awards at a show is one way to promote the animals to visitors at the barn. Awards, especially ribbons, should be displayed out of reach of the hogs.

6

THE HOG BUSINESS

In the pages ahead, I will endeavor to discuss the hog business and how the animals earned their reputation as mortgage lifters on farms all across the United States.

However, before I offer small-business strategies, it's important to reiterate that pork production as described in this book has been threatened in the recent past. As touched upon in chapter 1, at this writing there are a number of swine megaoperations dotting the rural landscape. These farms breed hundreds of sows and raise thousands of growing hogs. Only about 20 percent of all hogs are raised outside confinement today.

For decades, it has been predicted that agribusiness corporations would gobble up pork production the way they have both egg and broiler production. To present the obvious, however, hogs aren't chickens: They take much longer to reach a marketable weight, and they're still not being bred with anything near the uniformity found in broilers. Hogs also eat more than chickens, need more space, produce more manure, and require more labor.

Some of the giant swine operations already have begun to falter and fall into bankruptcy; as a group their growth curve has begun to slow, and there are hog industry observers who believe that any investment into such production will be a short-lived one. Indeed, many of these units seem to be starting up with a short-term recovery plan on investment; the thinking is that they will be forced out of business early in this century by humane-treatment, labor, and environmental issues. The waste lagoons of these facilities have been liabilities for decades, the buildings have no other uses once the technology outdates them, and energy costs and growing air pollution are now also a factor in the demise of these hog factories.

Twenty years ago, there existed a role for specialists who raised nothing but hogs on small to midsized farms. Actually, there were even more specialized roles within that hog-raiser category: Not only were there folks raising hogs, there were feeder-pig producers, hog feeders/finishers, seed- and show-stock producers, and even a few experimental breeders. It was probably too specialized an approach to the production of a single marketable agricultural item (a **commodity**) to endure. It isn't wise to rely on a commodity that depends almost entirely on the whims of a wholesale market with a demonstrated desire to reward naught but production in volume.

Today, although fewer small producers exist, the public has begun to look for healthy,

These are good-size shoats for a range situation. They should also be receiving a full feeding of a good growing ration to ensure desired growth and performance.

humane, and environmentally friendly alternatives to eating corporate pork. Smaller producers now are taking a shot at a more artisanal approach to pork production, returning it to many of its historical roots and making of pork something more akin to fine wine than to chicken nuggets.

Running a Manageable Operation

All the reasons to raise hogs that I've cited before remain valid, but these days hogs are best used as part of a truly diversified production plan. When you raise hogs aiming to target a few different markets and do so alongside several other sets of modest-sized, money-making undertakings that do not compete for the farmer's resources, you'll find that they will add real earning power and economic security to the small farm. You might be managing the production of 10 to 20 sows (or even fewer) in a venture that nets from $2,000 to $8,000 annually and serves a niche market, for example. Or say you have five sows producing F_1 gilts and feeder pigs or butchers for whole-hog sausage. (F_1 means the first-generation offspring of a cross between animals from two separate pure breeds; such offspring have hybrid vigor.)

Either of these hog ventures would be a good fit with a few beef cows and a pasture poultry operation or a pedigreed sheep flock. If we use hogs like our great-grandparents did, as one piece of a diverse farming enterprise,

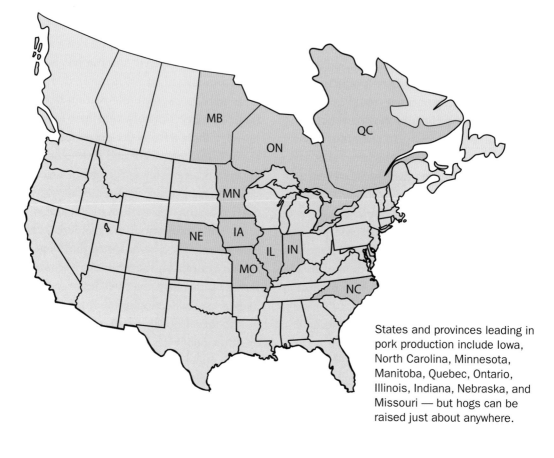

States and provinces leading in pork production include Iowa, North Carolina, Minnesota, Manitoba, Quebec, Ontario, Illinois, Indiana, Nebraska, and Missouri — but hogs can be raised just about anywhere.

their value really shines, whether the family farm is a 1st- or 10th-generation operation and whether it is located in Maine or Hawaii.

Big hog operations certainly seemed initially to get more than their fair share of positive press attention before the ills of the business began to be reported, and although this positive press is on the decline, much is made of their so-called **efficiency of scale** — that is, the fact that they are able to tend more hogs with basic operational methods, which presumably pares costs. However, nowhere has that efficiency of scale ever been actually documented. Some believe it may not appear until you own sows by the thousands — if then. The big operations have to maintain some semblance of cash flow to satisfy their investors. They are thus locked into producing × amount of hogs per diem, regardless of what is happening in either feed or hog markets.

A hog enterprise can fit into the day-to-day mix of life on a great many small farms without taking labor or resources away from other home ventures, and the farmers needn't focus their attention solely on hog-production volume each day the way factory managers must. Sows can be bred to farrow in any of the 12 months of the year so as not to interfere with the schedules of other undertakings. Many hog facilities can be made to do double duty by housing other species (a one-sow farrowing house with a solid floor is also a dandy place to brood baby chicks or waterfowl), and you can fine-tune the management of five sows far more easily than you can that of five hundred.

See the appendix for a sample of a small-operation calendar year in which farm business ventures do not compete for the farm's and farmer's resources.

GETTING INTO THE HOG BUSINESS

If you want to go into the hog business, the first point to resolve is why? What do you hope to achieve by raising hogs for the market? A great many small-scale producers make the move to swine production after a few years of feeding out a few pigs for the table needs of family and friends. They have gotten past the myths and misunderstandings about hogs and hog production and found they honestly like working with the porcine species. They often begin with a gilt or two that had initially been headed to the freezer.

Have Options

A litter of pigs can easily number from 8 to 12 at weaning, and with that many porcine mouths to feed, you must have a clear vision of their intended use and where they are to be sold. Don't take it for granted that you'll always have a wide variety of outlets for your production. Nor can you assume that a niche market now taking 20 head yearly at a fine premium will want or can absorb another 20 head.

We have always lived in Missouri, a part of the Corn Belt, and we have always had two of the corn industry's "Big I's" for neighbors: Iowa and Illinois. Still, for a time when I was first home from college, a single large hog-feeder operation — just one person buying feeder pigs and growing them to market weight — would often dominate and dictate the feeder-pig market in our three-county area of east-central Missouri. This individual would sometimes buy 150 to 350 pigs weekly and have upward of two thousand head on feed. Some seasons, this would be the entire week's supply of pigs at one or even two local pig auctions. It would also define the market for several weeks thereafter. Without this individual in the market, feeder-pig prices fell by as much as 20 cents per pound in a week.

This points out the danger of relying entirely upon a single outlet or a single marketing option. The traditional venture for small-scale swine producers has been feeder-pig production. Space and feed requirements are fairly modest, production can be quickly built up to a level that affords at least a monthly paycheck, and sweat equity can replace a great many capital goods (money, equipment, fancy housing) in the production of feeder pigs. Alas, the market for feeder pigs may be the most volatile that exists for any agricultural commodity; prices move up and down quite dramatically and often in just a very few days, and this market breaks before any other in the hog sector.

Define Your Turf

It is up to you as a small producer to define your own turf and pursue a variety of markets. The stranglehold on the local market, caused by the individual running the large feeder operation I described earlier, was broken when a graded feeder-pig auction (pigs from different farms blended) was established in the region with the support of a number of small producers. Feeders from as far away as central Iowa began to patronize it.

At Willow Valley — far, far from a showplace — we maintained a herd of three to eight purebred Duroc and Mulefoot sows and gilts. From them, we sold an average of 1½ to 2 breeding boars per litter, along with a few breeding stock gilts and some feeder pigs, and we have established a whole-hog sausage/cured-meat business ourselves. Certainly, not every purebred pig born is a candidate for the breeding herd, and feeder-pig sales are a strong backup market for us at times. Even with small groups of pigs, we often get top dollar because our pigs have a reputation for good performance, and because many buyers are seeking the spare gilt pigs that are left in a group after we pull the keeper boars. When you raise reputable purebreds, even the seconds retain high value to buyers.

Breeding stock and sausage are our prime marketing options. We target the boars to other small producers in our area and encourage them to develop pork products for direct marketing in our local east-central Missouri farmers' markets — a group of small farmers working together at several outlets with direct sales to consumers.

This is the key to working with small numbers: turn output into premium-quality production, then work to add as much value as possible to it before initiating direct marketing yourself. You can do this by sorting carefully, testing for parasites, offering animals at prime weights, validating the herd health, and so on. Feeder pigs are a sort of safety net that can be sold nearly any week of the year at one of a number of nearby consignment auctions. Our emerging pork-sausage sales were not a market we originally foresaw, but it's one with every bit as much promise as breeding-stock sales.

How Many Hogs for Success?

A sort of no-man's-land for pork producers is emerging. Ironically, until a few years ago, most producers were told to aspire to a level of production that incorporated from 75 to 200 sows. But producers with this many sows have hit some very big walls.

Production at these levels can totally dominate a farm, tie up at least one person full time with the hogs, require extensive investments in rather specialized swine-raising facilities, and make the whole farm totally dependent on the whims of the swine market.

Smaller numbers actually help family farmers maintain their greatest strength, flexibility in the face of adverse circumstances. The 10-sow producer should have the time and space for a number of other small ventures that will give the farm greater economic stability and orderly earnings, fully employ available labor, and add to the environmental integrity of that particular farming unit.

What drives the hog industry is the butcher-hog market, and when it slumps, survival often hinges on the producer's ability to tread the troubled economic waters. The small producer has a number of options: He or she can sell a pig crop as feeder pigs to reduce losses, feed out the pigs if feeder prices are not satisfactory, or even sell out completely and pull the hog houses up around the barn to ride out an especially trying period of distressed prices. Most breeding herds would benefit from the trimming away of the least productive third of their members; 60 percent

> ### FINDING THE ADVANTAGE
>
> I am aware of a small producer in Maine — a state not normally associated with pork production — who was initially building a Midwestern-style, farrow-to-finish hog operation. He had been building sow numbers and adapting volume production measures until he hit a sharp downtrend in butcher-hog prices. A cutback in herd size was mandated, but he was in an area with a very limited potential for the marketing of cull sows.
>
> He turned the old adage "If you can't sell a commodity, you'd better be prepared to eat it" to his advantage. He began turning those sows into whole-hog sausage and marketing it directly to consumers. At last contact, he was trimming his sow herd to 20 females to produce butcher hogs for his steadily developing sausage market.

or more can be trimmed away without impairing the solid genetic base needed for a herd to continue.

Conversely, growth need not be driven by outside forces. Small producers can build to the level at which they are comfortable without causing competition for available resources with other enterprises on the family farm or business.

The small swine producer has to concede nothing to the big operators or to the segment of the farm media that is so supportive of corporate farms. Small producers have access to the same genetics and can still reach most of the same markets. In fact, there are a few niche markets that are uniquely their own. They can tap into the same supplies of feedstuffs, and they can access appropriate technology and data through many of the same outlets that serve the high-volume producers.

Unfortunately, small farmers are too often identified with the wrong kind of production. Several years ago I was brought up short while strolling through a local auction and sale barn by the sight of a pen of solid black barrows that were crowding 350 pounds in weight. This was also the first whole pen of No. 4 grade butcher hogs that I had ever seen in my life. These are really fat hogs, with carcass lengths of less than 28 inches, backfat toppings of 2 inches, and poor muscling.

Overfinished, small boned, overaged, and just plain wastey, they were the talk of the barn that day. Later in the day, I caught up with one of the auction owners, a man who saw and handled thousands of hogs each year, and asked him about those big black hogs. He recalled them immediately and told me they had come from a small farmer who sent a couple of bunches each year to that sale. The farmer kept poor genetics, fed them sloppily, and did no marketing — just dumped them when the feeders were empty. But he was a small-farm anomaly.

Think Quality

Such an auction was perhaps the poorest of all possible places to market butcher hogs. It took place in a small sale barn that didn't pay much attention to quality or promotion, with an audience made up of a handful of farmers and traders. Once again, small farmers had taken a direct hit on the quality question because the owner of those fat hogs was perpetuating a negative, stereotypical image. There were other small farmers who had exemplary hogs: We sold them, as did many other small farmers. Volume producers have the numbers, however, to draw buyers directly to them.

As the meat markets of this nation become more and more quality driven, the price premium for quality will exist for those who can deliver the desired hogs, rather than hogs by the tractor-trailer or pickup load. When hogs don't pass muster on quality, the best thing to do is get them off the farm, take your lumps, and do a better job next time.

With quality in place, your next task is to maintain it at optimum levels of production, the operative term here being "optimum" rather than "maximum." Toward this end, you must shop carefully, invest correctly, and execute all management activities in a truly timely fashion.

Comparison Shop

Small farmers lack clout in the marketplace more often because they think they have convinced themselves that they have no clout than because of the actual conditions there. Comparison shopping for inputs, seasonal buying of inputs such as feed grains, and bulk buying and the like can make for substantial cash savings for even the smallest of operations.

Two elevators or grain dealers in the same town may be as much as 5 to 8 cents apart on a bushel of corn and $1 or more apart on a 50-pound sack of protein supplement. One feed dealer may try to cultivate small accounts by offering services such as inexpensive deliveries, while another may offer an extensive inventory of animal health and specialty products.

BE A SAVVY SHOPPER

Numerous university studies of swine producers and their farm records have shown that producers who are top ranked in their abilities to keep costs down often pay nearly one-third less for hog inputs, which include feed, breeding stock, equipment, and supplies. They use reserves to buy during the off-season and seek discounts for the volume of their purchases and the ability to use cash. They do a better job of shopping around and are more current in their knowledge of the marketplace than producers ranked below them. In addition, they tend to show increases in efficiency and output. They are succeeding not because of their greater numbers of hogs but simply because they are doing a better job. I believe this is true in part because they begin with an emphasis on quality. (See "Think Quality" on page 128.) They don't stint on inputs to the breeding herd or feedlot group. Good hogs don't eat any more than the other kind and often — very often — they eat less.

Most years, feed-grain prices are lowest at harvest. Sixty bushels of corn can be stored in a bin made from the hopper of a discarded 60-bushel self-feeder; 150 bushels will fit into a gravity wagon box set in the bay of a barn; an old chest-type freezer cabinet can hold as much as 1,000 pounds of shelled corn; and a 55-gallon drum will hold 300-plus pounds of the same grain. On-farm storage makes a pay-ahead and store-ahead plan a cost-cutting mechanism for even the smallest of country holdings.

A somewhat opposing school of thought that can also make a good statistical case for itself holds that a small producer should buy feed needs weekly, bimonthly, or monthly. In this way, the producer can average out the market highs and lows of the year. It's the way we do it at Willow Valley.

Know Your Market

The successful producer is very attuned and aware. The good ones are well read. At Willow Valley, at least six different swine-oriented publications arrive here monthly, and more

are available. Good producers are also plugged into the market outlets that are available to them on a regular basis — not just when they have something to sell — and they are in regular contact with their fellow producers. They study not just the movements in all the markets that concern pork production but also what causes those movements to occur, such as numbers building, drought, world affairs, and the like. They maintain detailed enterprise records and use those records to guide them in making management decisions. They know what it costs to produce a 40-pound feeder pig, a 100-pound butcher hog, or a service-ready young boar.

The good pork producer, then, is a person with both a plan of action and a specific purpose or purposes for his or her operation. It's a person who farms ideas first with a pencil and paper; if the ideas pan out there, then — and only then — does that producer invest sweat equity and precious capital into them.

TRAITS OF THE SUCCESSFUL PRODUCER

It's no mystery why some producers are successful. They share several traits in common. They know and honestly like hogs. They also have a clear plan for hog production, emphasize quality over quantity, and have multiple marketing outlets for their hogs. They develop and enhance these latter three qualities by doing the following:

· Reading hog-industry publications

· Staying "plugged in" with market outlets

· Keeping in touch with other producers

· Knowing what's influencing hog markets

· Maintaining detailed enterprise records

· Using their records to make management decisions

A Shift in Feeder-Pig Production

The feeder-pig trade as it was conducted for most of my life virtually no longer exists. Ten years ago, feeder pigs could be sold at four different weekly auctions that were within an hour's drive from home. Only one exists now. Throughout the Midwest, most feeder pigs now are produced under contract and for transfer from farrowing quarters on one farm to hot nursery facilities on another. In other words, pigs are produced under contract for a set price if company genetics and practices are used. They then move to another for growing and another for finishing. They can generally be canceled at the whim of the corporate entity, such as Farmland or Premium Standard.

This may now be done with pigs as young as 19 days of age and still largely dependent on a liquid diet. This is done in part, I believe, to circumvent modern CAFO (Confined Animal Feeding Operation) regulations as to the number of hogs that can be kept in a confinement unit as an environmental safeguard, which is based on a regular reporting schedule. Pigs younger than 10 days of age normally do not count in these population caps.

In the mid-'90s, feeder-pig market outlets faded rapidly and for a number of reasons. As butcher-hog prices plummeted, many tried to dump early weaned and hothouse nursery pigs onto available markets, and those pigs just did not work in the traditional feeder-pig marketing and rearing system. I can recall numerous accounts of large lots of such pigs being driven into the auction ring and drawing not a single bid. Some were given away there, and some could not find a taker even when offered for free.

With the great number of pigs now under contract, there is actually a shortage of available feeder pigs in many areas. However, there is no longer a market for large droves of feeders to go from a farmer/farrower to a farmer/feeder. Driving the small-producers market now are those wishing a few head to feed out for family and friends; for show; or for producing pork for one or more specialty, niche markets.

These markets exist far outside any traditional structure. Most of the sales are small-lot exchanges between the producer and the person who will be the end consumer or the one to be dealing directly with that person. The net result may be that this is one of the very few truly supply-and-demand-driven markets left in all of production agriculture.

Once or twice a month, a small producer phones in on our local radio station's noon swap program to sell a litter or two of feeder pigs. For roughly the last two years, his price has stood at $1 per pound for 50-pound pigs regardless of what the butcher-hog market is doing. And he gets it because he is nearly the only game in town — one of only a handful in a five-county area with such pigs to sell.

True Supply and Demand

A recent phone call from a college student in Nebraska revealed that her family is regularly marketing 55-pound purebred Berkshire shoats for a most impressive $2 per pound. They are sold in small lots to a local market of buyers wanting a very distinct kind of pork. These buyers want animals developed in a natural way, and they are willing to pay a premium price for animals that meet their expectations.

None of these producers are selling pigs by the hundreds of head, as was once the norm in the feeder-pig trade, but they do have a swine-based venture making a healthy contribution to their farm's bottom line and one that can be changed in its focus to meet the needs of another niche market with little or no disruption in its basic structure and the farm as a whole.

The key to success in this market is to be really plugged in and know what makes the market tick. Successful producers know their buyers; know what they want; and know why, how, and when they want it. Those buying feeder pigs for their own table use, for example, generally want animals that will reach a handy harvest weight by the time fall temperatures are beginning to cool. Those marketing pork to others may need feeders to harvest during barbecue season or feeders of a certain breeding background that will appeal to restaurant chefs and others in upscale markets.

This is currently not a market for which to gear up heavily, but the producer will probably want at least one or two groups of two to three sows, each bred to farrow at the same time. This should give numbers adequate to meet the needs of a number of small-lot buyers or a couple that wants 10 to 12 head for a special venture. These small lots are worth an added premium for the extra handling, time required in marketing, and their increasing unavailability.

Show Pigs

Another side to the feeder-pig trade and a bit of the seedstock trade, too, is the big and still-growing show-pig market. These are pigs for market-hog show — to be grown out and sold for harvest, often as a part of the show. Raising for the show-pig market is an outlet made for the small yet tightly run operation. We touched on this in an earlier chapter, but the simple truth is that show-pig production has now largely supplanted breeding-stock production for a large number of purebred-hog producers.

A top show-pig candidate now may bring 10 to 20 dollars a pound for a 40- to 50-pounder.

And there is a market that is nearly as strong for gilts with the potential to produce top pigs and the boar semen to sire them.

It is a venture not for the faint of heart, however. By one account, the members of a nearby FFA chapter spent an average of $100 per female to get gilts and sows bred artificially for spring litters for their county fair and the Missouri State Fair. This fee got them a breeding certificate (needed for registering the offspring) and two sticks of semen (enough for two services, roughly 12 hours apart and beginning 24 hours into standing heat).

There are those who tinker with show-pig production in much the same way an earlier generation tinkered to create hot rods out of family sedans. There are lots of twists and turns in type; some of them look good, and some are so extreme as to create animals that are totally impractical. For a time, there was a very real concern that the show-pig sector might become something apart and completely different from the other arms of pork production.

Hog shows have changed greatly in my lifetime. During my senior year in high school, the general hog show classes at our county fair began with the breeding-stock classes at eight a.m., but the market hog classes — the shortest part of the whole show — wouldn't commence until late afternoon. Now the market-hog classes dominate; they may extend for several hours, even at the local level. The market-hog shows may follow one or two jackpotted shows held earlier in the season, and very few young people now exhibit breeding animals. Thus, the show-pig market is primarily for barrows and gilts that will meet a handy market weight for the show or shows in which a young person is interested in participating.

A smaller, secondary show-pig market exists for purebred and crossbred females that can be bred to produce show pigs themselves. These are generally targeted toward older 4-H and FFA youngsters and those wishing to enter into show-pig production themselves.

A New (Old) Type of Pig

Much has changed in the two decades since I first wrote this book, and one of the most exciting changes for smallholders can be found in the trendy show-pig market. Producers are now witnessing a major shift in show-pig type: Buyers and judges are less attracted to pigs with extreme musculature and leanness and more interested in an animal that makes more sense from the farmer's standpoint. Attention is now focused on thicker, deeper-sided animals that should prove to be more sound and productive in the real world.

When an animal is taken to too great extremes in one area, overall performance can come to suffer. Buyers, judges, and raisers all understand this, and that understanding accounts for the shift in the trend. With excessive leanness — a hallmark of the corporate hog that gave us "the other white meat" — came problems with stress, durability, and reproduction. Those buying gilts of the winning type in the show-ring found them to be total wrecks in the farrowing house, if they could even get them bred. This shift in attention to more traditional hog types makes good sense. These are supposed to be market-hog shows, and that means honoring animals that are examples of profitability and productivity, not hogs that will win in the quarter mile.

It is imperative that those engaged in this venture stay absolutely current. This can mean considerable travel to shows, auctions, and judging events, but attendance can engender in a raiser the level of expertise that is very much a part of this pursuit. These producers must really study pedigrees, know emerging type trends, know the thinking of show judges who are most often working in their areas, have good communication skills, and be able to work well with young people.

Calendar Calculations

The show-pig market requires that producers be seasonally equipped. For example, in the Midwest and Northeast, most state and county fairs occur between June and September. In the Southwest and South, where summer temps really climb, state fairs and a number of major livestock events take place in the fall and even winter months, so producers need to time their breeding and farrowing accordingly.

A good show pig generally reaches 230 to 260 pounds, a good show weight, at 160 to 180 days of age. Keep in mind that this may be slowed a bit by hot weather and that the super-lean animals seem to take a bit longer to reach those desired weights. In my high school days, we showed a lot of hogs in the 220-to-240-pound weight range and had some of the better ones getting there by 5 months of age.

Aiming for those lighter weights — down to a 200-pound cutoff at some shows — some March-farrowed pigs were big enough to show at the later county fairs and the state fair. April pigs were right for the American Royal show in late fall in Kansas City, the first of the big shows of the winter season. And yet now, there is a real suspicion that some December-and-earlier-farrowed pigs are making their way into early-season county fairs here. Most of our Missouri fair shows were meant to be for "spring" farrowed pigs; that is, pigs farrowed in January, February, and March.

Over the years I have seen a lot of fudging on birth dates, and it was once common practice to trim the heads on older barrows to make them look "younger," but the good animals produced through straightforward practices generally shine through anyway.

When bred for optimal productivity and cash flow, the sow herd will produce two pig crops per year. One, generally the first litter produced in the year, will contain the pigs of the right age for many of the local shows and events. Some producers, marketing nationally, may get their second pig crop (born roughly six months later) sold, in part, into an area with a different show season. And to that end, you now see producers in different regions sometimes working together to create a greater number of selling opportunities throughout the year and to keep down the costs of seedstock inputs.

Out of that second crop, the producer may be more likely to sell breeding gilts to produce show pigs, a few breeding boars, and the rest as butcher stock. This can be a very high-cost marketing niche; however, much hinges on the marketing process, and the pigs have to prove competitive in the show-ring.

A sow herd for this market will be a rather unique creation. Three or four sows can produce a fair number of pigs suitable for one or two nearby county fairs. At our local fair, a little more than a hundred youngsters fit and show two to three hundred hogs each year. The numbers are reflective of the popularity of this project nationally, but at this particular show each youngster is allowed to sell one animal in the exhibitors' auction that follows the conclusion of the livestock shows. At last year's fair the hogs, lambs, calves, rabbits, and chickens had a combined selling price of more than $300,000.

CAN'T FOOL MOTHER NATURE — OR FARMERS

A book could and probably should be written on dubious show-ring practices, but there is still no better arbiter of good livestock type than head-to-head competition in the show-ring. It is validated by the fact that it is open competition under the hard eye of all manner of veteran producers gathered at ringside and on farms all over the country, where animals defined in the ring will ultimately have to produce. When it does function this way, wrong-headed type trends soon crash and burn, and the real nature of the animal is never overlooked or abused.

Mix 'Em Up for the Show-Ring

In our area, one of the largest herds focusing on the show-pig market maintains slightly fewer than 20 sows of five different breeds. Artificial insemination is used to produce both purebred and crossbred litters, and a Hampshire boar is kept for cleanup and natural service. The reasons for this great variety are many and range from practical to inspirational. Here are some that have been shared with me over the years:

- Variety sells. Some want purebreds, others want crosses for a bit of added hybrid vigor, and some people want a black pig for Bobby and a white one for Susie just to keep down the conflict and confusion.
- Dark-colored hogs show up better under the lights at a night show, and white hogs show best in the daylight.
- The crossbreds generally have a heavy dose of Hampshire breeding for the desired erect ears, black coloring, growth, and tighter meat type.
- Some minor breeds have traditional roles in Midwestern show-rings and are valued highly in other regions, even if their popularity has diminished in the last few years. One breeder I know works with a few of these — Chester Whites, Black Polands, and Spotteds — and though he wouldn't admit it, I suspect that, like me, he remembers better days for those breeds.
- The Duroc breed is used more for a crossbreeding base than to produce purebred show animals, although good red hogs are real showstoppers when they're driven out.
- The crossbreds are, most often, a first-generation (F_1) cross of two different purebred parents.

When founding a show-pig herd, select females carefully for good mothering and solid, middle-of-the-road type. The extreme females seldom make good mothers and are often hard to get bred in a timely fashion. I have seen several producers buy some of the more extreme, higher-placing gilts at local fairs, with the idea of breeding more show winners from them. Wrong-o!

Buying winning gilts for the purpose of mothering show pigs rarely works. Doing this ignores one of the basic tenets of a market-hog show: that these shows should be considered terminal events. The animals at a market-hog show are bred for one reason and one reason only: They are to be harvested as high-quality meat animals at as early an age as possible. Their growth and meat type should come from a carefully planned mating, and the boar selected to sire them will have much to say about this.

The smallholder exploring this market will most likely do so with a small set of sows bred pure or a handful of F_1 litters from Yorkshire, Chester, or Duroc females bred to a Hampshire or Berkshire boar of good show type. And the best foot in the door is to just get out there and rack up a few wins. It won't take the kids long to find you when that happens.

Boar in a Bottle

I am no fan of artificial insemination, although it is being embraced more and more by this whole seedstock sector and by high-volume commercial producers. It can certainly extend the use of a good boar and better help to disburse his costs, but it can also rapidly narrow a breed's gene pool, kill seedstock sales for all producers, and reduce producer control of this most important of inputs — and I am not really sure of all of its supposed health-management benefits.

PROBLEMS WITH ARTIFICIAL INSEMINATION (AI)

Artificial insemination (AI) is far from a natural process. With its use come many concerns. Among them are the following:

- Estrus is not always easy to detect, and some females show no signs of heat. There is nothing better at detecting a female in estrus than an intact male, and his mere presence may even induce estrus onset.

- It is a practice with a limited level of success. Successful AI breeding rates can be as low as 35 percent of exposed females bred, and anything with a 65 percent success rate on the first service is considered very good. Meanwhile, the rate for natural insemination is generally 90 percent or greater.

- There is usually a quite limited time frame in which to get a female bred for a specific show or event. Miss a breeding time with AI, and you're done for three weeks.

- This technology makes possible the use of a popular boar but relatively few of his sons, to the point where whole breed populations can become badly skewed. The boar in a bottle can be used everywhere and for a long time. In England a few years ago, one Holstein bull had 50,000 descendants on the ground due to AI. That narrows gene pools terribly.

A modest number of boar studs have grown up to serve the show-pig market, and they now supply both fresh and frozen semen all across the United States. They have created a demand for a handful of boars from a few select breeds and have bid up some state fair winners into the low six figures in recent years; this at a time when good purebred boars can be had for much less than $400 per head.

In many areas, there are now suppliers and artificial insemination technicians that can supply semen from males of a number of different breeds and then do the actual insemination work. This service is often sold as a package.

Farrow-to-Finish Production

The most common swine-production option pursued in the United States is the complete route, commonly termed **farrow-to-finish** production. It is more costly than feeder-pig production but has been demonstrated to be the most consistently profitable type of swine production.

The farrow-to-finish operator owns the pigs from birth to slaughter. This process means that the producer must invest in an additional set of growing/finishing facilities, the hogs will be on the farm for an additional three to four

Weanling or lightweight feeder pig

A young pig soon approaching feeder pig weight

Raising pigs from farrow to finish means taking them from birth to slaughter.

A market-weight hog

months, and the growing hogs will consume substantial amounts of feed grains.

The small-farm farrow-to-finish operator normally works with just one or two sets of 4 to 10 sows each. With the sow groups bred to farrow at 60-day intervals, they can easily be worked through a single set of farrowing huts, and farrowing in really severe weather can often be avoided entirely.

Some might choose to feed out one set of pigs and then market the next as feeders or even split-market each farrowing. A pen of growing/finishing hogs can really tax a small farm's resources, more so than other farming ventures. To reach a market weight of 230 pounds, each will consume $50 or more in feedstuffs over and above what it received as a young pig and will require an investment in labor of at least 1½ hours to market, along with a rather specialized set of facilities.

A number of small finishing units can be bought or built on runners or old mobile-home frames to hold 25 to 65 head of finishing hogs. They will represent an investment of $2,200 to $4,600, not including feeding and watering equipment. These units save a great deal of space on the small farm and will qualify for investment credits under current tax laws.

Finishing Hogs

Finishing used to be called "fattening hogs" before we hog producers got image conscious, and with the hogs of 50 or 60 years ago, it was a quite literal description of the process. Buying feeder pigs and taking them to a good market weight is a costly venture with a substantial amount of risk. It is commonly pursued by high-volume grain producers who are feeding hogs to have one more grain-marketing option.

Finishing feeder pigs is a far-less-common venture among small farmers, but many pursue it on a small scale as a part of a niche-marketing venture. They may feed out small numbers of hogs for sale directly as processed-pork products, such as whole-hog sausage, custom-fed freezer pork, or hogs for roasting, or for sale to other specialty outlets.

As a small-scale finisher, you can often find bargain pigs by buying small or irregular groups of feeder pigs that pass through auctions in the wake of the bigger, more desirable droves. If you're equipped with adequate facilities for housing them, you can have some real bargains by buying lightweight pigs in very cold or inclement weather. One veterinarian in our area paid for a good part of his school expenses by buying and repairing ruptured pigs — that is, pigs with a hernia of the umbilicus or scrotal variety — along with pigs with downed ears, boar pigs, slop pigs, and the like. Although this is not necessarily recommended for farmers who are not skilled in performing medical treatments, he was able to provide health care at cost and then feed up the low-cost pigs to good slaughter weight.

Breeding-Stock Production

Producing breeding stock, which you'll learn a lot more about in the chapters ahead, is also an option of the small-scale producer. Contrary to popular belief, you do not need sows by the hundreds to produce boars — which are still largely sold just one or two head at a time. A few years back, two of the three division winners in the Hampshire show at the Missouri State Fair came from a single Hampshire sow herd with just five sows to its credit.

If they are of good type and bred to an equally good boar, a group of five females can easily produce 8 to 10 breeding-quality boars to sell twice a year, along with a goodly number of quality siblings that can be sold in other ways. Those same five females could also produce two 10- to 15-member groups of F_1 crossbred replacement gilts annually. (F_1 means the first-generation offspring of a cross between animals from two separate pure breeds; such offspring have hybrid vigor.)

Over the years, we have owned a number of sows that regularly produced two to three keeper boars per litter. Even a quick reading of breeder sale catalogs will show sows producing as many as six marketable boars in a single litter. We once owned a Hampshire sow that produced 35 marketable boars during the five years she was in our herd.

Purebreds

In the swine sector both purebred and crossbred breeding animals have substantial value. With purebreds the primary market is for boars to be sold to commercial producers, who use them in crosses to produce feeder pigs or butcher hogs. They are typically sold as 8- to 10-month-old boars ready for light breeding service.

The farm press is filled with stories of pedigreed boars selling for four and five figures at events like the World Pork Expo or summer conferences, but the greatest number of boars now changes hands in the $200- to $600-per-head price range. Still, that is at least twice the price of a butcher hog, which costs about the same to grow out.

A few years back, when small, commercial sow herds were kept in a more varied state, there was a bit more demand for purebred gilts, and some people held that the real profits in a purebred operation stemmed from gilt sales. This may well be the case with some pure breeds, now that the gilts have better profit potential if sold at feeder-pig weights as youth-project animals. Gilts from the white breeds are used by some commercial producers in what are called parent and grandparent matings. They are used to produce in-house F_1 and F_2 herd replacements for the sow herd.

Crossbreds

Crossbreds — F_1s, to be precise — have value as both breeding boars and replacement gilts. The boars carry a blend of breed strengths in addition to packing that always desirable hybrid vigor. They are perceived to be more hardy and vigorous in the breeding pen than their purebred counterparts. The F_1 gilts can be bred to a boar of a third breed not present in their cross and produce crossbred pigs with good hybrid vigor and reasonably predictable benefits, and their own gilts can be brought back into the breeding herd without disrupting breeding patterns and heterosis (hybrid vigor) levels.

Commercial gilts normally change hands at 250 to 300 pounds and at 6 to 8 months of age (when they are ready to breed or close to it). They typically sell for a $25- to $100-per-head premium over and above what they would bring as No. 1 butcher hogs. Their price base is generally pulled from one of the stronger Midwestern markets, such as Omaha or Sioux City. The boars generally sell in the lower range of purebred-boar prices, with most going for between $250 and $400 per head. For some producers they are the Chevrolet alternative to the purebred Cadillac. Without good purebreds, however, they would not be possible, so there is a place for a good purebred boar in even the smallest commercial swine herd.

POPULAR CROSSBREDS

A new generation of crossbreds that are finding favor are the ¾/¼ crosses, such as the now very popular ¾ Hampshire/ ¼ Duroc boars that result from the cross of an F_1 Hampshire × Duroc female to a purebred Hampshire male. They have a color pattern similar to off-belt, purebred Hampshires and tend to have a very high degree of performance predictability.

On the gilt side, the most popular crosses are the Hampshire × Yorkshire, Hampshire × Landrace, Hampshire × Chester White, Duroc × Yorkshire, and Hampshire × Duroc females. Bred back to a colored male of a third breed, they should produce market hogs that grow well and yield good quality.

With such crosses, white breeding can be introduced into a herd or reinforced without tipping the scales too greatly away from the meat-type genetics that predominate with the use of the colored breeds.

At Willow Valley, we have sold a number of Duroc × Hampshire boars over the years and found them to be nearly as popular as our purebred Duroc boars. Still, they were at best a compromise between the two breeds and not as strong in the traits most commonly identified with their parent breeds as purebred representatives of those breeds would have been.

Breeding-Stock Producers

Breeding-stock production takes place at various levels. It is rather like a three-tiered pyramid.

At the top is a rather small group of producers who shape most of the trends in breeding-stock types and market primarily to other purebred producers. They dictate type and trend because they invest the time and effort to develop, show, and promote their type developments.

The second layer consists of producers sometimes called **multiplier breeders**. They may work with multiple breeds, purebreds and crosses, or all crosses. They are volume producers seeking to sell to the larger commercial producers. They can supply 100 gilts for repopulation of an entire herd or 20 boars that can work together in a group. Some of these producers may still show hogs, but their primary marketing tool is performance data, which they use to formulate estimated-breeding, growth-rate, and feed-efficiency values for the breeding animals they sell. These estimated values allegedly make it possible to project to what level a trait can be improved through a specific animal. A given male might thus be able to shave one day off the existing herd's average days to market.

The larger the tested populations, and the more testing that is done, the more precise these estimated projections will be. Much hinges on the exactness of the testing procedures, the integrity of the test data, and the size of the test population (the larger the size, the better the data). To me, this procedure seems to lack adequate safeguards, is costly to carry out, and, like much of the effort that went into amassing testing data in the past, often does little more than discourage or drive out small and midsized family-farm breeders.

A third group pursues purebred production and use on a more traditional level. These producers may raise just a single breed of swine, and they quite often operate with 50 or fewer sows. They are the ones helping to maintain a number of the minor and endangered swine breeds, along with the other traditions of the family farm. Their primary market lies within 50 to 75 miles of the front gate, they sell a lot of $200 to $300 animals, they produce purebreds but without record pedigrees, and they sell hogs in small numbers to other small and midsized producers.

Many producers feel strongly that the pork industry was sold out by its leadership when breed associations led the charge toward factory farming and confinement operations. Because these breeding associations bought into the "bigger is better" way of thinking, many producers of even popular breeds have opted out of using them and now do their own breeding documentation.

We do not record pedigrees at Willow Valley because we believe in purebred breeding but not the politics that surround the purebred animal. Pedigrees are only as good as the people who hand them to you.

Direct-Marketing Your Own Pork

Pork production once meant that the hog raiser would turn his or her production over to the processing and retail sectors for workup and marketing to the end consumers. It was a concept that worked fairly well until agribusiness turned pork into another commodity

CENSUS NUMBERS: MORE WOMEN ON FARMS

Traditionally, the farm wife and older children played a key role in caring for the farmyard and its denizens. Census data continues to show growing participation by both women and young people in the management and care of livestock.

Reasons for this include the modern reality of both marriage partners working off-farm for part of each day, so that all family members must share livestock care. Women also take the lead in providing value-added measures and direct-marketing. Many women have become effective communicators as to the value of modern pork.

Women's presence in animal agriculture is not going unnoticed by the support sector, either. Feedstuffs and bedding materials are being marketed in smaller, more manageable bags; tractors of 40 hp and less are nearly always shown being driven by women; 4-H girls are choosing hog projects; and young women now make up the larger segment of most class years in veterinary school.

At farm conferences I've noticed more and more couples in attendance, along with women that are sole proprietors. They generally ask thoughtful questions and seem more open to change and to the value of direct marketing.

In our area alone, I know a daughter/father partnership producing show pigs, a mother and daughter offering pork products from their multigenerational family farm at farmers' markets, women leading and assisting with youth livestock projects, and many couples where the wife manages the farm while the husband works away from home. Homemaking and farm tending are, as they always have been, the work for both genders.

like turpentine or paving stone. They want it in tidy-sized pieces and cheap, cheap, cheap!

Numerous health scares, environmental horror stories, and a desire for animals to be treated humanely have caused an ever-growing segment of our population to want to buy their food directly from farmers and to have more input into how it is produced. And fortunately, they are nearly always willing to pay a premium for this, to have "their farmer," and to be able to interface with him or her.

It is a potentially quite lucrative market but one that depends greatly on the producer's people skills and ability to manage a potentially highly perishable product. Pork is one of those "sell it or smell it" items of production. And while today's consumers are quite receptive to farm-fresh pork, most of these good folks are many generations away from any type of farm experience. A pile of pork is a bonanza they may not be well equipped to handle, and a great deal of consumer education is in order.

Educating the Public

The first big lesson is that a hog is not all hams and chops. One warm afternoon not long after this book was first published, I received a most interesting phone call. The gentleman informed me that he was calling from high up in a skyscraper in New York City and he had a business proposition he wished to discuss.

He had a plan to market the type of cured hams he so vividly remembered from his youthful vacations in Virginia. I agreed that such hams were indeed the stuff of legend, as are cured pork products from many parts of the country. I was able to answer his many questions, including how to acquire a good supply of fresh hams, two each from the many thousands of hogs still produced in our part of the country. But my first question brought his entire plan to a screeching halt! What was he going to do with the rest of the hog? I asked.

The idea of having a large supply of fresh pork on hand certainly has great appeal, especially when the wheels on the economy begin to wobble a bit, as they did late in the first decade of the twenty-first century. The problems begin, however, when consumers learn that in that ham-and-ribs supply is included feet, tail, ears, head, and jowl; that to get tenderloins you have to give up pork chops; and that curing costs extra.

We sold our first hog directly to consumers nearly 35 years ago. They were suburban neighbors of my sister and were, as Dad said, green as a gourd. It took a month or so before we had any hogs ready to market, and the result was long weeks of handholding for a couple who thought their neighbor's cocker spaniel was a big animal.

Over and over they told us how thick they wanted their pork chops. We told them no, we wouldn't know how many pounds of sausage there would be until the hog was worked up. We fielded all of their questions—and they had quite a few. But it wasn't all bad; we found an element of satisfaction in their interest and in telling them the story of our farm and hogs. It also gave us a far better understanding of and appreciation for the modern consumer with limited experience.

Self-Education

Good farmers cannot be expected to be butchers or full-service retailers, but for this particular venture some of the strengths of these professionals must be added to the skill set of the hog producer. When farmers and smallholders venture into direct marketing, they find, especially in the early going, that substantial blocks of their time will have to be given over to the marketing process. They must listen to consumers' wants and then seek out their answers. By some estimates 4 out of every 10 hours given over to a venture relying on direct marketing must be given over to the marketing process.

This can be a great drawback to some, as they feel they do not have the right skills nor the desire to give over time for this end. A Saturday morning at the local farmers' market takes 8 to 10 hours out of a week for preparation and actual time on site. And the clientele, especially the inexperienced and naive, can be most exasperating with the nature and frequency of their questions. To those accustomed to selling their production by the trailer or truckload, small-lot sales don't jibe with their image of what a farm marketer should be. You may become exasperated, but if you don't listen you will be broke, too.

Most local markets for pork can be quite finite in nature, but they do pay well, often very well; that is, if your product is what they really want. Keep in mind that part of what consumers want is some of your time. They respect you as the expert raiser and want to tap into your knowledge about how best to buy, keep, and prepare the meat. And they look to be respected in return — to be treated as more than bipedal ATM machines. They want to feel that they are just as important to you as you are to them — and they are!

But Can I Make a Living?

Profit margins for a direct-market meat animal can run from $100 to $200 or more per head. I have regularly achieved this with hogs marketed as whole-hog sausage at our local farmers' market. Others are marketing to chefs, selling through CSAs, and marketing through specialty stores and other collective efforts. Yet too many farm folk, when they hear these numbers, want to know where and how quickly they can deliver five hundred to a thousand head every month. That ain't gonna happen.

Many buyers are basically niche markets, capable of absorbing just 10 to 30 head per year. The simple truth is that high-volume pork production feeds almost entirely into a wholesale market. In such a market a profit of $5 per head would be considered quite good. For 20 head sold with a $100-per-head profit margin, the net returns are $2,000 per year. Couple that with nine other ventures of a similar nature, and you have a well-diversified small farm earning $20,000 per year.

At the $5-per-head margin, it takes four hundred head to produce that same amount of return — doing with 24 sows what the niche markets can do with 2. And making that kind of money with far fewer animals means far less time spent and investment put at risk.

Pricing. The most widely known pork products are the hams, loins, and bacons and the shoulders processed into roasts or steaks. Once these cuts are processed, however, a lot of pork is left on the processor's table. Producers should consider that there are added costs if some of these products are cured before marketing and that a whole ham can often be a daunting purchase for some of today's smaller families. As Dorothy Parker is believed to have said, "Eternity is a ham and two people."

The sale of these most familiar cuts of meat also may require the producer to figure in added refrigerated storage and time in the marketing process. Pricing must reflect these factors and the added processing costs also. With these products it is probably best to be led by consumer requests, have them processed only to fill existing orders, and require an advance payment from the consumer. Very few will back away from an initial deposit, and it should be large enough to cover any added costs the producer is apt to encounter during processing. As much as I like cured ham, I like it much better when someone else has paid for it; curing is an added cost that the consumer should know about and pay for.

Many producers begin in this market by selling whole and half hogs and delivering them to a local processor that the producer and buyer have mutually agreed upon. This inspected processor returns ready-to-market, frozen stocks to the grower. The cost of the delivery is generally then included in the price the purchaser pays for the hog. The consumer typically buys the animal by the pound based on the animal's live, on-foot weight or the animal's **hanging weight** — what the carcass weighs following slaughter, skinning or scraping, and evisceration. This can give the buyer a better idea of meat to be received and its quality, but hanging-weight prices are generally higher than liveweight prices.

Added value. How to price these animals is always a bit of a challenge, but generally most enhancing measures add 3 cents to 10 cents per pound in value over the base market on the day of delivery or when the deal is struck. For example, high-percentage Berkshire or Tamworth breeding adds a 10 cents-per-pound liveweight premium. Antibiotic- or additive-free feeding should add a similar premium. Humane rearing practices, selling a single animal, delivery, feeding to a special harvest weight, and other production measures shown as free-range production should all add value to the consumer. And never sell a half carcass unless you have a place to go with the other half or plan on using it yourself. Remember: Don't get greedy. Your goal should be to build repeat customers and not prey on the unsuspecting.

Working with a processor. The key to all of this is access to a nearby processor who does custom work. They are still fairly common throughout the Midwest and in many other locales. If you have trouble finding one, ask your deer-hunting friends where they have their deer processed.

When the animal is delivered to the processing plant, the operator generally has a form for the buyer to fill out. It should outline the costs of various processing measures and allow the buyer to check off his or her needs. These will include the curing of the hams, bacon, or jowl, if desired; the desired thickness of steaks and chops, fresh ham or ham steaks (if not cured); scraps and odd pieces worked up for sausage or pork burger; any specialty or ethnic-themed products (brats instead of bulk sausage, for example); size of freezer packages; payment details; and more.

Added costs and options for the buyer. Added costs may include a kill fee, yardage (the cost to sell above commission), and any specialty charges. Some processors request a security deposit from the buyer, and all fees will be due in full when the meat is picked up. The meat will generally be quick frozen, and the fresh pork can often be picked up within a week of the animal's being delivered to the plant. Cured and specialty sausage products may take an added one to two weeks and require a second trip for the buyer. I have always found it best for the buyer to communicate directly with the processor specifics such as how thick they want the pork chops, how many to include in a package, whether they want their sausage spicy or mild, and so on.

Whole-Hog Products

If selling whole and half hogs to businesses and individuals and using a processor as a middleman is not your cup of tea, marketing a product that utilizes the whole hog may be more to your liking. Two examples of this are whole-hog pork sausage and a relatively new product known as pork burger.

Pork sausage is a staple of the classic American breakfast, and in the Midwest it is often served as a part of the evening meal with fried potatoes and apples. Here it is an autumn tradition. Sausage is enthusiastically accepted all across the country. **Whole-hog sausage** is just that: the hams, loins, bacons, and shoulders all go into the mix. The odd items like the head can also be boned out and included. The only item not used in the sausage making will be the pork ribs; thus it is a most efficient, frugal use of a butcher hog. Because the primal cuts just mentioned are in the sausage mix, it is often in the 80 percent or greater lean range and thus comparable to the finest ground-beef products. Modern pork sausage can generally be processed into a bulk product packaged in 1- or 2-pound sticks or in brat-sized links. Few small processors are equipped to make the small breakfast links.

Pork burger was developed in the 1970s, just to the north of us in Pike County, Missouri, along the Mississippi River. It was created as an option for those wanting a ground-pork product with a bit more versatility than sausage. A noted Bowling Green, Missouri, locker

---

WHOLE-HOG SAUSAGE MARKETING TIPS

Choose only a very mild seasoning blend for the bulk product. Those wanting a hotter or spicier sausage can then season it themselves. The classic sausage seasoning, sage, actually doesn't freeze very well and can impart a bitter flavor to the meat in frozen storage.

Have a variety of offerings. In our area, we now see the brat-sized sausage links seasoned in a number of different ways, including Italian, Greek, and hot. Here, too, a mildly seasoned product is the best choice to appeal to as wide a number as possible, but it is nice to have seasoning options that can help draw various ethnic markets.

Offer 1-pound samplers. Especially in the early going, 1-pound sizes will be the best packaging choice. These appeal to first-time buyers who wish to give a food item a modest trial before committing to it in any substantial way. Also keep in mind today's smaller families; 2-pound sticks will probably be as large as they will ever go.

---

facility developed a special seasoning blend that imparted to ground pork a mild flavor and cooking qualities that compared favorably to hamburger. It was first offered at a number of local and regional pork- and swine-themed events, and its popularity spread through the Midwest and then outward.

There is little shrinkage in the pan, it works well on the grill, it can replace hamburger in most recipes, and for me it makes an especially savory cheeseburger. From local processors it is available as a bulk product and as 4-, 6-, and 8-ounce burgers.

These whole-hog products have a good shelf life, can be stored easily in a home freezer, have modest processing costs, and allow the fullest possible utilization of the entire pork carcass. And everything that adds a premium price to hams or chops can be applied to these two products also. They can be derived from additive-free and humanely raised hogs, can be from a range-based system of raising, and are a natural fit for those wanting to make the most efficient use possible of surplus animals from a rare breed–raising program.

Marketing a processed product will require a bit more of the producer, and he or she will have plenty of homework to do. The following are some of the necessary steps you'll need to take to legally and safely deliver your product from the business-plan stage to that of packaged meat in the buyer's hands.

1. Research all state and local food-processing regulations, specifically as they relate to your farm and pursuits. This information should be available from your state's Department of Agriculture and your county's office of health or sanitation. Many feel most comfortable making these inquiries through a third party such as a county Extension agent or farmers' market officer.

2. If your product is approved by a USDA inspector at the local processor, it can then, generally, be sold by the producer within the borders of that state. USDA-inspected meat is legal for sale in all states. State-inspected meats are for sale in-state only unless an adjoining state has reciprocity. Do be sure that the processor does not use packaging materials that normally go to end consumers and that the packaging does not have a "Not for Resale" stamp or sticker. It should carry a USDA inspection logo, however.

3. You will receive the product flash or fast frozen from the processor in the desired forms and packaging. Personalized packaging can be acquired, but there are very exacting laws as to content labeling, handling directions, and weight documentation to which you must adhere. Our local processor uses a very attractive blue and white sleeve for sticks of bulk product and a clear tight wrap for links.

4. Transport the delivered product quickly to your frozen storage in basic, insulated coolers.

5. When taking the product to selling events, such as farmers' markets, the two most common means of transport and handling are well-iced coolers or a small freezer unit that can be plugged in on-site after a short time in transport.

6. Transport the product in small lots that will sell rapidly, and do not attempt to refreeze any product that has begun to thaw. It will be safe to eat as long as some ice crystals remain in the meat.

7. Maintain a supply on hand no larger than can be sold easily in 30 to 60 days.

Pricing for your processed meat should cover all costs, including the time and expense given over to the marketing process. A base price might begin with that which is being charged for a comparable product in local stores. With that established you can then begin adding a modest premium for special touches you have included. The more rurally situated your market outlets, the less likely you will be to charge top-dollar prices for your production. Those prices are to be had only close to or within urban centers.

Most consumer surveys show a willingness on the part of consumers to pay a price premium of 5 to 12 percent for products they feel are most in keeping with their values. Purchase of this kind of item is not just an ordinary restocking of staples; it is also a feel-good purchase and not an empty one. The consumer expresses his or her values by supporting the business and raising practices of the producer. This is what is at the heart of successful direct marketing and the reason many consumers have turned their backs on the corporate raisers. When consumers support these farmers, they are giving their personal stamp of approval on the farmer's relationship to the animals and the environment, the quality of the animal's life, and the quality of the meat product.

What about Organic?

As more and more of the public have begun to decry agribusiness practices and turn away from corporate products, small producers are finding increasingly specific niche markets to serve. In fact, fulfillment of certain unique products has changed the way some smallholders have raised their hogs and produced their pork for decades. Others have always raised their hogs in ways that are pleasing to the current public taste, but some production methods are easier to employ than others.

For example, many farmers are trying their hand at filling customer demand for organic meat. While I am not at odds with this type of production, I do have some very long-held concerns about its practicality. For example, I don't believe in giving the whole herd a round of vaccinations every time one animal sneezes. But to speed healing; ease suffering; and maintain a valuable, sometimes endangered animal, I do believe antibiotics have a role to play and that giving antibiotics can be a caring and sound management choice.

GENETICALLY MODIFIED CROPS

Crops that have been genetically altered are characterized as **genetically modified organisms (GMO)**. Few crops in the United States have been as extensively modified by genetically altered production as yellow corn. The process has been used to counter insect resistance and is now being employed to extend drought tolerance. It is also used in one of the major breeding lines to improve rooting strength in many hybrid seed varieties.

Strong evidence suggests that beneficial insects are harmed by Bt corn (*Bacillus thuringiensis*) and that some gene modifications have escaped to wild populations of corn's parent plants. These modifications have been found in Mexican *teosinte* (parent corn plant) nearly a mile above sea level, although GMO producers swore up and down the pollen wouldn't climb.

Many are reporting palatability and performance problems with GMO corn varieties. But there is no long-term research data as yet on the consequences of this practice.

At the time of this writing, ethanol demand has corn futures flirting with the $4.20-a-bushel mark, and organic corn, when and where it can be had, is at or above $8.50 a bushel. Organic protein supplements are equally high. For the farmer hoping to raise organic swine, you can add to these pricing hurdles the fact that one of the primary breeding lines of corn used to fix good rooting traits in numerous hybrid varieties is now known to be contaminated by GMOs (see box on page 147, Genetically Modified Crops).

If you live anywhere in farm country where the wind blows and rain falls, you are very apt to be compromised by wind drift and runoff from your nonorganic farming neighbors. For example, rainwater runoff from crops recently sprayed with pesticides will flow onto any nearby land that is downslope, and wind can move that spray hundreds of yards. Because one can't control the weather or a neighbor's farming practices, attempts at organic production can sometimes be futile.

Organic production is a valid pursuit, but only as long as it can be made cost-effective. Right now comparable selling prices to those of organic products can be had with additive-free, humanely reared, and locally produced pork and hogs and at much less cost to the individual producers.

PAINFUL PATH TO ORGANIC PRODUCTION

Organic pork production now hinges on access to a very upscale market outlet, as costs to produce can be quite high. A bushel of organic corn can, for example, cost three to four times as much as a conventionally produced bushel of corn.

Farms producing organic livestock face rather rigid rules and are subject to inspection and quasi licensing, and only a limited number of health-care products and practices are permitted. Animals treated with antibiotics may have to be completely removed from the farm.

Challenges to organic production now include the growing use of genetic-modification practices in the production of many hybrid-corn varieties. Producers in heavily farmed areas are going to have problems with water runoff and wind drift that bring chemicals onto their farms.

The decision to opt for organic production must be carefully thought out, as it is a decision that will affect the producer and his or her farm and herd for many years to come. A plan of limited antibiotic use, outdoor rearing, feeding less intensively produced feedstuffs, and selecting for more vigorous natural type may be a better alternative for many.

It can take years to get a farm certified for organic production and can be quite expensive, so these producers need rapid access to populations of consumers with substantial disposable incomes. But perhaps most prohibitive is the fact that the organic producer must be able to ensure the absolute purity of the production of his organic products because of the rather serious health problems that can result for people who unwittingly consume impurities.

Contented Hogs

Pork from hogs raised in an outdoor system, fed additive-free rations, and selectively bred for natural hardiness and vigor will certainly satisfy most of the desires and concerns of today's more informed and demanding consumers. Hogs from a small population and an open-air system will certainly benefit from a more natural, healthful, and less stressful environment. Instead of building on that "other white meat" ad campaign, modern producers should look further back in both advertising and agricultural history and strive to produce pork "from contented hogs." It is excellent!

Pork has always been one of the most distinctive and flavorsome meats in the human diet. We now know that some breeds of hog do produce a unique, tastier pork than others. The breeds that currently, verifiably fit this bill are the Berkshire, Chester White, Tamworth, and Duroc. The meat they produce (especially that of the Berkshire) will appear different. On the plate, the flavor and eating qualities will be different, too.

Berkshire pork, for example, will appear darker, has fat flecking within the muscling (marbling), and its pH has even been proven to be different. Some of these differences can be quite subtle, but they prove that the foods we eat are more than a mere source of calories. Many people — including such diverse groups as those in the Slow Food movement and Asians seeking a richer, more sustainable, locally based diet — see in these differences much of value and validation of their beliefs and good efforts. One of the highest compliments I was ever paid as a farmer came from a high-volume producer in our area who said he bought our Duroc boars to produce the pork his family ate.

This is a market niche that the beef industry has long focused on and is the basis for its "Certified" Hereford and Angus beef programs. Until recently, few if any swine breeds were as fixed in the minds of consumers as are these two great old-line beef breeds. However, today, in areas around the world, including the lucrative Pacific Rim, buyers are starting to ask for Berkshire pork and pay a real premium for it.

More lab work, public sampling, and taste testing by chefs and the press must be carried out with the pork from these and other swine breeds. I suspect that pork from the Black Poland, the Gloucester Old Spot, and the Hereford may eventually qualify for inclusion in this delectable group. The word about the mouthwatering qualities of these heirloom breeds will spread, and as that happens, I believe we will see swelling interest in them, not unlike what's happened with the many heirloom and endangered poultry breeds.

SUCCULENT HEIRLOOM PORK FOR SALE

In the production of heirloom pork is a marketing window that needs to be thrown open wide and from which the farmers involved must shout the virtues of their chosen breeds. Eat and preserve these breeds, you ask? Yes! Not every pig born is of breeder quality; substantial numbers of surplus males are produced in every pig crop. Sales of heritage-breed meat animals generate the funds needed to keep family farms in place and those animals breeding into the future.

Lean *and* Green

An issue of some concern to small swine producers is the ever-growing market weight for butcher hogs. Forty years ago, a 220-pounder was a prime animal. Today the prime animal is one that ranges between 245 and 260 pounds. Butcher hogs are grown as large as 300 pounds and are no longer penalized with much lower prices per pound as they once were. The point is often made in defense of high-volume producers that we are producing as much pork now as in 1950 but with far fewer pork producers.

Actually, many heavier hogs are being produced now, and the technology that is supposed to be utilized by top producers also enables mediocre producers to do a mediocre job with a lot more hogs. With housing, mechanization, and a friendly banker, the producer who once had a hundred-sow herd can do the same lackluster job with a thousand sows. Big tractors don't necessarily make good stockpeople. Even with improved genetics, bigger hogs are older hogs — their growth curve has slowed, and they are putting on fat cover instead of building muscle.

Big hogs do pare costs of operating on the killing floor and the processing line, which makes them a more profitable hog to process, but there is little about them conducive to better pork. They stand around longer in the feedyard, feed efficiency goes down as size increases, and even an average of a pound or two of average added gain can move over the marketplace like a crushing wave. These big hogs may be on the farms 10 to 20 days longer than the lighter-weight butchers of a few years ago.

In many countries known globally for the quality of their pork, younger, smaller hogs go to market. Forty years ago, you could have yourself a 220-pound butcher when the animal was just a bit older than 5 months. They were efficient gainers because they were young and growing muscle, and they moved through finishing facilities and off the farm in short order. And the producers didn't fill up storage facilities with large, harder-to-market primal cuts.

I believe the younger, lighter-weight market hog may again have a role to play with those engaged in the direct marketing of pork and butcher hogs. They are the porcine equivalent of the Cornish game hen or the "baby chicken" broiler now showing up in pricier restaurants. And this approach would certainly produce the basic cuts of pork now in keeping with today's smaller families and the on-the-go trend toward fewer large meals eaten at home.

Selling the public on the virtues of eating younger pigs will be a bit of a marketing challenge at first, but think of the appeal of a naturally lean animal that has made optimal growth on modest amounts of grain and protein feedstuffs. This is "green" production and a "green" pork product at a time when that word has enlivened the marketplace as few others ever have.

LOOKING AHEAD

Before we proceed to the chapters on acquiring and managing breeding stock, let's take a look at the future. The breeding-stock industry now relies almost entirely on the Duroc, Yorkshire, and Hampshire breeds and to a lesser extent on the Landrace. Many believe that the remaining pure breeds and their backers will have to follow the lead of the sheep industry, where a large number of pure breeds have been taken up by true family farmers who maintain small breeding groups as part of truly diversified farming programs. They support a number of breed groups and shows, are very supportive of each other, promote purebreds even as market animals, and sell their breeding animals to other small producers.

If this continues to happen, there will be fewer very-high-priced hogs in the future, but as the breeds are cultivated for their individual strengths, they will reaffirm their rightful roles as building blocks within the pork industry. The trend toward gourmet pork, such as what is now growing up around purebred and high-percentage-bred Berkshire market animals, may continue. The texture and cooking qualities of the meat of these animals are highly esteemed in Asia, and they are now beginning to be recognized here.

Check the roots of most of the great and lasting purebred herds and flocks in this country and abroad, and you will find that they extend back to just one or two foundation females. This points up that while it is a challenge to launch a lasting purebred operation, it is also doable on a modest, truly human scale.

We bought our first purebred gilt for $117.50 at an auction conducted by local FFA youngsters. Our first purebred boar was farrowed in the stripped-out hulk of a 1953 Plymouth. When we bought him, we were driving a 1952 Plymouth. On the way home Dad began laughing — it was the first time he had ever known of a hog having to come down in the world.

As in nearly every endeavor in life, agricultural or otherwise, start small, grow as you learn, emphasize quality, and treat everyone in exactly the same way that you like to be treated. An unwritten but ironclad rule in the production of swine breeding stock is, "Sell only the kind you like to buy."

7

PREPARING FOR YOUR HERD

I've always believed in having a place to put a hog before you go about acquiring one. It makes life so much simpler, and it keeps your boots and jeans oh so much cleaner.

To house a sow herd and a breeding boar, you will need two sets of facilities. One will be used to contain the sows during breeding and gestation, the other is for farrowing and pig rearing. Your housing and fencing choices will hinge on the approach you take to swine production.

On the small farm, sows are generally maintained in drylots, on pasture, or in a combination of the two. At Willow Valley we use drylots to contain our Duroc and Mulefoot sows. The ground is hilly, and in a normal year we receive 35 inches of rain or a bit more; the sloping lots drain quickly and reduce problems with mud. In flat lots, huge mudholes can form during rainy seasons, especially around feeding and watering equipment.

The Drylot

A **drylot** is a contained outdoor space on the ground, used continuously, with all feedstuffs provided. We try to provide at least 250 square feet of space per sow. On flatter terrain or where rain amounts are greater, sows may need two or three times that amount of lot space. The drylot system lends itself especially well to hilly parcels, wooded lots, and those little odd areas not convenient or suitable for cropping or grazing. Along wooded fencerows, field margins, and rolling parcels just going to scrub growth, hogs can neatly fit in, be sheltered from sun by the shade that trees provide, and yield a good economic return on a parcel that might otherwise produce no income.

One of our drylots was in continuous use for nearly 10 years and still sheltered sows in a safe and healthful environment. However, many producers prefer to rotate drylots every few years, plowing them up and sowing them to grass or legumes for a time. Leaving them idle for 12 months will disrupt the life cycles of a great many harmful parasites and disease organisms.

The Pasture

Few sights are prettier than a set of hogs, slick and shiny, on a rich, green pasture. It is a picture of health and wholesomeness. Hogs can fare well even on total legume pastures — something few ruminants can do.

A set of crossbred gilts fast approaching breeding age

ENVIRONMENTAL SAFEGUARD

At the foot or bottom end of each drylot, we maintain a strip of sod 10 to 20 feet wide. This naturally filters the runoff from the lots and thus serves as an environmental safeguard. A hard rain clears the lots to what is essentially a new surface, and the sod maintains the lots in a natural way, with minimal runoff getting past the strip and little or nothing in the way of erosion in the lot.

Following the very wet years of 1993, 1994, and 1995, we idled one sow lot; within 60 days it had naturally sprouted a number of plant varieties — something quite exceptional, as many veteran swine producers will concede. Wastes both in drylots and on pasture percolate through the ground, and grass is a natural filter. On the smaller farms I know, most manure and urine is returned to the land in used or "spent" hay bedding. It is strewn onto crop ground and pastures in early spring before tillage.

Keeping hogs on a drylot can yield a good
economic return on a parcel of land that
might otherwise produce no income.

Hogs are not the most efficient users of pasturage, however. They are omnivores and have a single gut. They do not totally utilize the browse — twigs, leaves, and shoots — and they need richer sources of energy. One-quarter acre of pasture will adequately carry up to four sows with their litters, but it has to be considered little more than a dessert option for those sows. Nursing sows raised on pasture must be given a full-fed lactation ration. Gestating sows on good pasturage might have their rations trimmed by ¼ to ½ pound of protein supplement and 1 to 2 pounds of corn daily. You need to closely monitor sows for tone and fleshing — see that they are satisfactorily gaining weight following weaning, whether they are on pasture or not.

During the last one-third of the pregnancy, sows on pasture should be fed virtually the same as if they were in a drylot. One acre of good legume pasture will supply something on the order of one-quarter of the nutritional needs of 10 head of growing/finishing hogs. Finishing hogs on pasture is probably not the most efficient use of today's higher-priced grazing and farming lands.

The Central Hub

A relatively new approach to rearing hogs on pasture entails maintaining them in a series of pens arranged in a wagon-wheel pattern off a central hub. Access points to the pens are at the hub, as is the primary drinking-water source. Also maintained at the central hub

Finishing hogs on pasture may not be the most efficient use of land in areas where grazing and farming acreage is high-priced, but it can still be profitable — and it sure is a pretty sight.

SWINE PASTURE CROPS

The pasture crop that would absolutely stand up to the wear and tear of grazing hogs would be a natural variety that has a great deal in common with Astroturf. Alas, such a species does not exist. Still, hardiness and durability are important factors to take into consideration when selecting plantings for swine pastures. Some of the hardier alfalfa varieties, clovers, and various grasses can be used to pasture swine. Multiple varieties create a more enduring pasture; mixes also are often less costly to establish and maintain than pure stands, especially pure stands of some of the legumes.

Hardier species of alfalfa and clover can be mixed in with various grasses when raising hogs on pasture.

are feed-storage facilities, handling and loading facilities, and cold-weather farrowing and nursery units.

This approach was used a couple of decades ago in the Missouri Ozarks, when this area was the feeder-pig capital of the United States. Breeding, much of farrowing, and growout took place in the rolling, wooded lots. Herding dogs were very valuable in these operations.

The wagon-wheel system greatly simplifies caring for hogs on range, reduces the need for costly service and access roads and lanes, and makes it possible to extend a single water source into a number of lots and pastures.

Around buildings and in areas where hogs are to be worked or sorted, I prefer panels, gates, or woven wire supported by posts set on 8-foot centers. The fencing material should be placed inside the fence posts, so there is no risk that the animals will push it off the posts by rubbing against it.

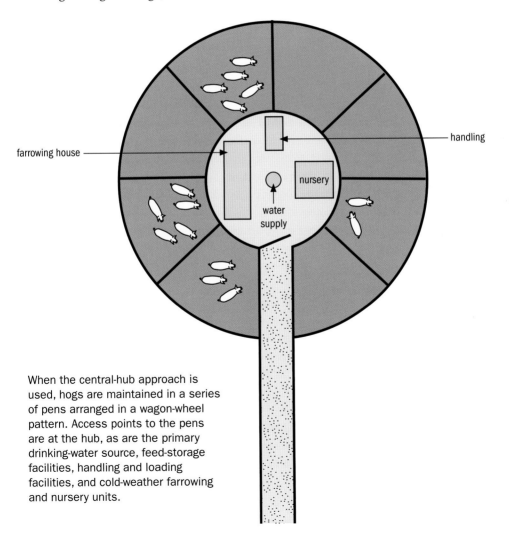

farrowing house

handling

nursery

water supply

When the central-hub approach is used, hogs are maintained in a series of pens arranged in a wagon-wheel pattern. Access points to the pens are at the hub, as are the primary drinking-water source, feed-storage facilities, handling and loading facilities, and cold-weather farrowing and nursery units.

Protecting Pastures

Hogs on pasture should be kept rung to minimize rooting damage to pasture crops and the soil surface. Such activity can kill valuable plant species and set the stage for erosion problems. **Ringing** is the clipping of soft metal rings to the top rim of the hog's nose or across the end of the nose with what is called a "humane ring." The ring causes mild pain to the animal when it tries to root and thus discourages rooting activity. The rings are inexpensive, simple to attach with a plierlike tool, and cause no lasting pain or disfigurement, and the animals can eat and sneeze. On the other hand, they are also easily lost (especially those in the tip of the nose), over time they may have to be reapplied, application requires the hog to be restrained, and there is the pain factor, which is troubling to some people. Still, there is very little in the way of alternatives to ringing other than raising the hogs on concrete.

Hoof action can also take a toll on soil surfaces. Feeding and watering equipment should be placed on runnered platforms or concrete pads. The runnered platforms, made from durable and inexpensive native hardwood such as oak, can be towed about to prevent

PASTURE OR DRYLOT?

To many, the idea of raising hogs on pasture seems as dated as high-button shoes or buggy whips. Yet in many parts of the Midwest where crop yields are legendary and land prices regularly top $2,000 per acre, pasture hog-rearing remains not just popular but also profitable.

In many counties of western Illinois, southeastern Iowa, and elsewhere, sows on pasture continue to thrive in the face of an industry seemingly bent on building ever-bigger, ever-more-costly buildings for the confinement rearing of swine. The pastures are a part of a successful and longstanding crop-rotation system; the hogs return manure, urine, and spent bedding to the land; and per-sow investments on these farms are measured in ten- and twenty-dollar bills rather than hundred-dollar bills. In a state-of-the-art confinement facility, the investment is often as many as 30 hundred-dollar bills per sow.

Rotating pastures after 12 to 18 months of use breaks up parasite and disease-organism life cycles. Tillage then eliminates any rooting or wallowing damage. Such farms thus have a margin of sustainability few others can ever hope to equal.

Hogs will take a toll on growing plants in pasture when they forage; the pointed shape of their hooves will damage the plants. To maintain a rather delicate pasture plant mix, you have to match appropriately the number of hogs with the available pasture.

A drylot will simply support greater numbers of hogs per acre. Hogs can also be held for longer in drylots before lot rotation is needed, and they are generally more accessible in a drylot. By using drylots, we have maintained five to eight healthful and contented sows on our 2.86-acre farm — something we could not have done in a full pasture situation.

For farms with less land and an allowance for grain, drylots may be the answer. For land-rich farms, hogs on pasture may be the best scenario, but keep in mind that they will need feed supplements.

Hogs on pasture should have a humane ring on their nose to lessen rooting damage to pasture crops and the soil surface.

problems with wallowing and mud buildup from developing. Placing the platforms or pads immediately adjacent to fence lines will make over-the-fence filling possible and greatly simplify feeding and watering chores.

Fencing Choices

Whether it's on drylots or pasture, the area you enclose for your hogs will be far greater than the few square feet needed for a growing animal or two on a slatted floor. Hogs must be contained in a manner that holds them safely but does not bust your budget.

I am from a part of the country and a tradition that holds that the very best fences are bull strong, baby-chick tight, and horse high. Unfortunately, such fences are now dearer than diamonds to erect. Further, hogs normally breach their enclosures by going under

them or, as Dad used to say, finding a little hole and worrying it into a big one. You are indeed trying to contain an animal with a bulldozer blade for a nose.

Hogs can be successfully contained with woven wire, a combination of woven and barbed wire, gate-type panels, or electric fencing.

Woven Wire

Woven wire in the 26- to-34-inch-high range is what is commonly called hog wire. It is high enough to discourage most hogs from jumping over it, is easily handled (the 26-inch rolls can be wound and unwound within the space between two rows of corn), is quite durable, and can be supported by wooden or steel fence posts. It is by far the best choice for perimeter and other permanent fences.

It is also the most costly among swine-fencing options—as well as the most complex and time consuming to erect. Setting woven-wire fences is one of the big jobs of our year, reserved for the summer months between planting and harvest.

Woven wire for swine fencing should be tautly stretched. Good corner posts set deeply (30 to 36 inches, to get below the frost line), braced in every direction from which the wire will be drawn, and selected for long life will ensure fencing that is of the greatest durability, safer to erect, and easier to maintain.

BARBED WIRE

Woven wire as a fencing medium can be enhanced and strengthened with the addition of four-point barbed wire. Positioning the woven wire 4 inches above the ground with a strand of barbed wire beneath it will create a longer-lasting fence line and one that will discourage hogs from rooting beneath it. Two more strands of barbed wire 4 and 8 inches above the woven wire will discourage jumping and make the fencing usable with other species of livestock as well.

Hogs can be fenced in with woven wire, a combination of woven and barbed wire, gate-type panels, or electric fencing. But woven wire, pictured here, is by far the best choice for perimeter and other permanent fences.

Hog Panels

Hog panels in some ways resemble woven wire but are made of much heavier gauge materials and measure 34 inches high by 16 feet long. Fastened to wooden or steel posts at 8-foot intervals, they make a durable, although rather expensive, fencing option. They can be attached to the posts with long-shanked staples, clips, or short lengths of tie wire. The panels can be taken down for reuse and can be handled by just one person.

These panels are also useful for forming simple gates and spanning water gaps. Wired up in such a fashion, they make what Dad used to call "pack gates." To open them you have to lift up one end and pack it around until the desired space in the fence line is opened. See an illustration of a hog panel on page 51.

Corner Posts

When I was a youngster, we used cedar posts that were 8 to 10 feet long and at least 8 inches across at the small end for corners. Dad liked to have his corner posts set in place for a year before attaching fencing to them, to be sure they would hold his fence wire fiddle-string tight. Posts left in for a while settle into the

earth and are sturdier; if you fence to posts too soon, they'll shift around, and you'll have more fence-maintenance work to do.

A Missouri fencing trademark was and is 8-foot-long crossties set 3 feet into the ground in a concrete footing for corner posts. I believe that some of those old creosote-stained monoliths will still be standing when the Gateway Arch of St. Louis is but a memory. Wooden corner posts should be at least 8 inches in diameter and pressure treated or cut from a long-lived wood such as red cedar or locust.

Double-bracing corner posts with treated poles or timbers will further strengthen their holding power. There is also a system now that makes it possible to double-brace 7-foot-long steel posts with other steel posts and use them for solidly anchored fence corners.

Line Posts

Line posts need to be set at 10- to 15-foot intervals, with the longest intervals used in the long, straight stretches. With frequent or sharp dips along the line, fence posts need to be placed closer together, to keep the fence pulled down to the proper height.

10- to 15-foot intervals

Line posts should be set at 10- to 15-foot intervals, with the longest intervals used in long, straight stretches. If you use both woven wire and barbed wire in the fencing, the pasture or lot can be used by other livestock.

Wooden line posts should be set at least 18 to 24 inches deep; they should be at least 4 inches in diameter. In among the wooden posts should be a few steel posts, to ground the fence line in the event of a lightning strike.

Five- to 6-foot-long steel posts can also be used as line posts. They should be set at the same intervals in the fence line as the wooden posts.

Electric Fencing

The fourth option for containing hogs is electric fencing — "**hot wire.**" It is the least costly fencing option, it goes up quickly and easily, and you can hang a couple of miles of it on the barn wall. It can safely contain hogs of all ages and sizes but is not a good choice for perimeter fencing because it is the most easily breached of all the fencing choices.

Chargers. A hot-wire fence receives its charge from a boxlike transformer commonly called a charger. There are chargers powered by electrical current, batteries, and solar power. Electric-fence chargers are one of the few things left in this world that operate for a few pennies a day.

Chargers need to be protected from the elements, which can be done either by installing them within an outbuilding or by covering them with some kind of weatherproof box or cover. I've seen chargers wired to a fence post, plugged into a $3 imported extension cord, and protected by nothing more than the bottom of a $2 Styrofoam cooler give years of service, but it sure ain't professional.

With the charger placed in a building, the charging wire can be passed through a hole bored in a wall and insulated with a short segment of rubber hose or tubing; run underground and insulated in a length of garden hose or conduit; or carried around and out of the building on ceramic insulators attached to structural elements and positioned well above head height.

The cases on most battery-charged models form weathertight containment for the controls and a six-volt dry-cell battery. In my experience, these batteries seldom provide maximum charging power for more than 60 to 90 days. Such batteries are not all that costly, but you do pay for the convenience of a charger that can operate far from the nearest electrical outlet.

Old six-volt car batteries that are still in good order can be used to propel these chargers. Each battery is generally positioned on a board on the ground beneath a charger; extra wire is used to reach from the battery's post

HOW TO GROUND CHARGERS PROPERLY

Battery- and solar-powered chargers can be set right at the head of the fence line, but what is critical to the successful operation of all three types of electric fences is having the chargers properly grounded. A 6-foot-long copper rod driven at least 4 feet into the ground will serve admirably to ground a fence charger. The rod must be driven deep into the ground to ensure that it is constantly in contact with soil moisture. In some areas a 6-foot rod may have to be driven nearly its full length into the ground. In very dry weather or climes, you may find it necessary to regularly douse the area around the ground rod with several buckets of water.

to the battery couplers inside the charger. For short spans of wire, there is even a new generation of battery chargers that are powered with flashlight batteries and clip right to the fence wire.

The most expensive chargers are generally the solar-powered models. They can give dependable service even during moderate periods of cloudy weather and can carry electric-fence usage to the farthest reaches of the family farm.

Electric-fencing materials. A number of fencing-material options are available for use with modern electric-fence chargers. There is a mesh material that resembles woven wire but that can be folded and unfolded for rapid erection and takedown, ease of relocation, and simple storage. There are tapes and ribbons that combine plastics with fine wire fibers to create high-visibility fencing that can be spliced simply by tying ends together. There is even a light-gauge barbed wire that can be used with electric-fence chargers.

The most often used wire choice for electric fencing is 9- to 14-gauge smooth wire. It can be rolled and unrolled for repeated use, is durable, carries the electric charge well, and can be erected over virtually any terrain. It is commonly sold in spools that contain ¼ mile of wire each.

Posts for electric fencing include such options as hollow and solid fiberglass rods; short, smooth metal posts; and even homemade models that give old concrete rebar a second lease on life. Most are 42 to 60 inches in length and go into the ground easily with a few taps from a shop hammer. In many soil types, they can even be pushed into the ground by hand or by simply stepping on their flanges.

Electric-fence chargers are one of the few things you can operate for pennies per day.

The fence wire can actually be threaded through some of the fiberglass posts or attached to them with simple metal clips that resemble cotter keys. They are ungrounded, and the current simply flows through them, too. To the wooden and steel posts, a variety of plastic and porcelain insulators can be affixed in various ways; these insulate the charged wires from shorting out against the posts.

With metal posts, I prefer to use the plastic insulators that can be slid up and down without disconnecting the wire. This is an especially useful feature when electric fencing is being used to contain growing hogs or in areas where substantial snow buildup occurs.

A single strand of charged wire suspended about 12 inches above the ground will contain most hogs from about 80 pounds up to breeding animals. At this height, a hog will most often contact the charged wire across the bridge of its nose, a point that is quite effective in turning the animal back from the fence line. A second charged wire 4 inches above the ground will contain small pigs and young shoats.

A strand of charged wire suspended about 12 inches above the ground (the bottom strand in this photo) will contain most hogs about 80 pounds and up.

Erecting the electric fence. When erecting electric fencing, I like to use 6- or 7-foot-long steel posts that are driven into the ground until the flange is well covered for my corner posts. I use porcelain doughnut insulators attached to the posts with heavy-gauge wire to anchor the line wires to the corners. Then, regardless of ultimate pen shape, I can run the wire in straight lines from one anchoring post to another. At the corners I join the line runs with short pieces of wire wrapped around the two lines in a coil-like manner. This ensures very taut fence lines and is quick and simple to erect.

The Range House

The basic unit for housing growing and breeding hogs on pasture and drylots is the range house. It's the Chevy of hog housing, and it can be used in drylots or pasture. Simply put, it is a three-sided wooden or sheet-metal house on treated-lumber or native-hardwood runners with an insulated roof.

Basic Design

Generally, the range house has no floor and varies in size from 8 × 10 feet to 8 × 16 feet or larger. These houses are positioned with the open end facing south or east, away from the prevailing winds in most of the United States.

The traditional approach to range-house construction was to leave one of the wide sides open and extend the roof overhang 18 to 24 inches out over the opened side. The overhang was a house's only protection from blowing snow and rain. The design saved some building materials, and with such a wide opening, the hogs entered and exited with little or no strain on doorways and structural elements. But that wide-open side was also an Achilles' heel.

ELECTRIC FENCING: HELPFUL HINTS

I have used electric fencing for many years to contain a number of livestock species. Along the way I've picked up a few bits of lore and wisdom on how to operate and maintain the fence line and charger for best results:

- When charging multiple pens with a single charger, install simple, blade-type switches in the various pen lines to simplify repair work. Then to isolate a segment of the line or a pen to make it safe to repair or even dismantle, you merely throw a switch.

- Hogs that have never been exposed to charged wire tend to run right through it on their first encounter. To train them to respect hot wire, expose them to it inside a woven-wire- or panel-enclosed lot. Run a short strand through the pen, place a bit of feed adjacent to and beneath it, and they will soon learn to give it a wide berth.

- Monitor the charger and fence line several times a day. Walking the lines and checking the charger while doing chores is a good practice. We've owned sows that knew immediately when fence chargers went offline.

- Walk the lines completely following strong winds and storms. Weeds, grass that's damp and heavy, and limbs across the fence line can short it out totally or greatly reduce its ability to contain hogs.

- Prefer solid-state fence controllers to those with replaceable, plug-type chopper controls. They cost more but stay online better, and you don't have the often-repeated cost of replacing the chopper control. Likewise, the fences with the ability to shock through wet weeds and grasses are preferable.

- Invest in one of the simple testing tools for measuring the strength of the charge being carried by the fence line.

This device is used to test the strength of the charge in a strand of electric fencing.

- Hogs are often reluctant to cross areas where they know charged wire has been used, so be sure to install frequent gates or panels in electric fence lines. Where pens come together, set a panel in the line and then butt two more off it to form gates for the separate lots. You can then add additional panels in those corners to form working pens in those areas. In these small pens within bigger ones, the hogs will feel more secure and at ease, and they'll be easier to sort, handle, and load.

Despite the overhangs, rain and snow could sometimes blow all the way to the back of the shallow beds. The house was more open to chilling winds, too, and in cold weather a large portion of the open side had to be boarded up to help contain bedding and body heat. Further, with its long open side, this house was more cumbersome to position in or adjacent to a pen, especially a smaller pen.

Improved Range-House Design

The improved range-house design is a simple rethinking of its basic design. Instead of a long side, one of the narrow ends is left open to face south or east. The two runners remain the same length, but they are placed under the sidewalls rather than across the back and front of the house. Such a house can be easily towed into, through, and out of a pen.

Its real benefit, however, is the greater comfort quotient for the hogs it contains. With a 5-foot front opening falling to a 4-foot-high back wall over a length of 12 to 16 feet, a true microenvironment is created deep within the house. Bedding is kept dryer, chilling winds are less likely to penetrate to where the hogs lie, and body heat is better contained within such a house. Some ½-inch black board or other sheet insulation should be placed beneath the sheet-metal roofing. This prevents the sleeping hogs' breath from condensing on it in cold weather, then falling back down on the hogs and bedding and chilling them.

For size and logistics of movement, few pieces of hog equipment are as daunting as a range house. I recall attending one farm-dispersal sale where two really huge, 14 × 24-foot range houses were to be offered at auction. They would not fit on any of the commonly available trailers of that day — they had actually been built with no intention of ever removing them from that farm. They sold for less than many of the smaller houses offered that day, and the reason for that was best explained in the words of a farmer I stood next to in the crowd: "Ain't only one way to get them things home, boy. Just hook onto 'em and don't look back till ya get there," he said.

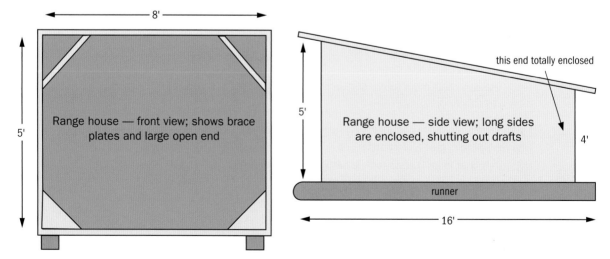

The newer type of range house has an improved design that features an opening on the narrow end, not the long.

Range-House Interiors

Inside the range house, each sow will need at least 12 to 16 square feet of sleeping space; an older boar may need up to 20 square feet. In cold weather, adding as little as 4 inches of clean straw bedding can raise comfort levels by an ambient level of as much as 10°F (5.5°C).

Range houses have a great many uses in the small-scale swine operation. One of our neighbors has a number of 8 × 16-foot houses he uses to shelter sows and litters while on range. When they are 1 to 2 weeks of age, depending on the weather and the time of year, he moves nursing litters and sows from their farrowing huts to wooded lots or pastures. He puts two or three sows and litters in each house, shutting them inside it with a hog panel wired across the front, and holds them there for a night to get them accustomed to their new quarters. He maintains the sows and litters in these houses until weaning.

In many places, an 8 × 10-foot (or a bit larger) house is used to make a portable nursery to hold the pigs from several litters following weaning. A solid floor of 2-inch-thick native-hardwood lumber or plywood is put into the house, and the front opening is reduced to a couple of small, pop-through doorways, which will admit only small amounts of potentially cold air to the sleeping area. The house is then pulled adjacent to a slatted-floor pen of equal dimensions. Feeding and watering equipment is set in the outside pen. Pigs of up to 40 pounds will need about 4 square feet of floor space each in such a unit.

MOVING LARGE HOG HOUSES

There are a few simple measures you can take that will render larger hog houses easier and safer to move about and situate. Here are some suggestions:

1. Bolt towing straps to both ends of each hog-house runner. Use U-shaped straps at least 1 inch wide and ⅜ inch thick, and bolt them onto the runners through holes bored back at least 1 foot from the runner ends. Any shallower and there is a very real risk of tearing them out during a very hard pull.

2. Brace corners carefully with 2-inch-thick hardwood lumber at floor level and plywood plates at the roofline.

3. Start a house moving with a slow and steady pull, not with a jerk.

4. Set the house on a 4-inch-high raised bed of tamped earth, cull lime, or gravel. This will keep runoff from passing through the house and the runners from settling deeply into the mud. A really old trick is to set each corner of the house down on a large fieldstone to keep the house from settling into the mud, also making it harder to move by towing.

5. Trench around the house to further direct runoff water away from the sleeping beds; keep the trenches cleaned out to prevent mud from building up around the runners.

Keep in mind that the longer a house sits in one place, the harder it will be to free it up and move it the next time.

Farrowing Structures

The other big structural requirement for those with a sow herd is a place for farrowing — somewhere to shelter the females during birth and for a few days afterward, until the pigs are off to a good start. In many months of the year, a sow can heap up a pile of leaves and old cornstalks to make a nest and raise at least a few pigs. But if you have litters coming at every season, you need housing that shelters the newborns from the elements, predators, and muck and mire.

Farrowing-house options range from simple floorless huts and hutches to one-sow houses with porch pens to pull-together or modular houses to centrally located, single-purpose farrowing houses. All work well in specific situations, and there is something to be said for and against each and every one.

At the core is the means by which the sow is contained during the time of farrowing and early pig rearing. Sows can be contained in farrowing crates, stalls, or single-sow houses (the latter is basically a four-sided version of the freestanding farrowing stall).

PORTABLE FARROWING HOUSE

The illustrations below show the framing on a portable farrowing house developed by Extension Service personnel at the University of Missouri in Columbia for Laclede County. It features a sliding roof, and it can be moved by a forklift on the back of a tractor. The forks slip between the skids, and the house is lifted by a hydraulic system. The house also can be pulled on skids, although in this case the skids should be sloped at the ends, similarly to skids for sleds. Notice that one side is for pigs and includes a pig guard, while the other side is for the sow.

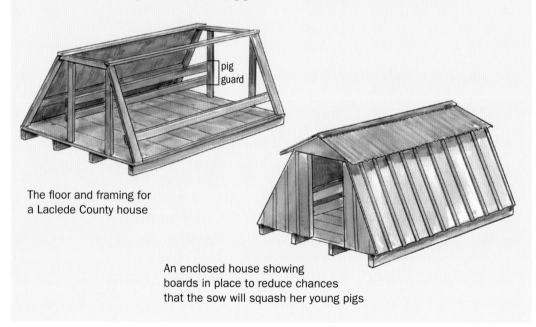

The floor and framing for a Laclede County house

An enclosed house showing boards in place to reduce chances that the sow will squash her young pigs

Choices in farrowing housing will be dictated by economics, climate, and enterprise goals:

- The producer selling boars and show pigs must be able to farrow in January, come what may. He or she will want to use draft-free housing with supplemental heat.
- The producer wanting to sell F_1 gilts or butcher hogs may be better served by a more seasonal approach to farrowing and the resulting cost savings. He or she will want to use floorless farrowing huts for each sow and litter.
- On some family farms, a combination of these options may be in order. The farmer will want to use some of both types of housing listed for the scenarios above.

Farrowing Crates

Even after decades of use, farrowing crates remain somewhat controversial, due to the constrictive way in which they contain sows. Crates do indeed hold the sows in close quarters — but for the express purpose of protecting very young pigs from overlay and crushing. During the first 96 hours following farrowing, death losses are greater than at any other time in the pig's life, and overlay is one of the primary causes of those deaths.

The basic crate is 5 feet wide by 7 feet long. The central part of the unit, where the sow is contained, is 2×7 feet. It is supposed to be snug enough to keep her from turning around and stepping on any of her baby pigs or flopping down on them when she lies down to nurse.

It holds the female in a rather exacting position, so that even the birth will occur in a specific portion of the crate. There the pigs will be easily accessible to the producer, supplemental heat can be directed on them, and they can easily find the sow's side and quickly begin nursing. A farrowing crate is best used for the 72 hours around birth, but sows should always be let out twice a day for feed and exercise.

Down each side of the sow's zone are 18-inch-wide by 7-foot-long pig bunks. Here the pigs can lie safely away from the sow, still have simple access to her side for nursing, are safely provided with supplemental heat, and can be proffered creep feed or milk replacer. These bunks are meant to be a safe retreat for the pigs away from the danger of overlay or misdirected maternal feet.

Initially, farrowing crates were made of 2-inch-thick planking arranged in a gatelike fashion with an 8-inch-high opening at the bottom for the pigs to have easy access to the sow when they wished to nurse. This crate design is still used in many areas due to its simplicity and modest cost; it can also be a homemade project using simple tools. However, over the years a number of modifications have been made to this design, and the great majority of crates are now constructed of steel pipe or square tubing.

Simple farrowing crates are built into some one-sow farrowing houses also. A Midwestern variation on the farrowing crate has metal-framed crates erected on 5×7-foot sheets of expanded metal flooring set on steel-pipe legs. The crate frames, including the pig bunks, are enclosed with plywood sheathing. They can be set up in various outbuildings for cold-weather farrowing or taken to the hedgerows and field margins for summer farrowing. Wastes fall through the mesh floors, and the crates are fairly simple to take down for cleaning or storage.

A narrower crate — 22 inches wide in the sow area — is often used to farrow gilts, which are both younger and smaller than sows. Another modification if space is at a premium is to provide a pig bunk down just one side of the farrowing crate. Some producers also install a nest box at one end of the farrowing crate to hold young pigs completely away from the sow's zone in a microenvironment all their own.

Crates do save young pigs, and with proper management, they can maintain sows in a comfortable and healthful manner. We have used crates made from native-oak lumber; while the number of deaths from overlay and crushing did go down, we also lost a few pigs to cuts caused by sows extending their hooves into the pig bunks.

Crate flooring. Nearly as oft discussed as crate design options is the flooring material that goes beneath the crates. On most small farms, crate flooring is still found in its most basic form — some sort of solid wooden flooring to keep drafts from coming up under the sow and her litter and chilling them. For us, native oak 2×6s or 2×8s laid side by side without slots provide both draft control and durability and are among the least costly flooring options.

There can be problems with wooden flooring under hogs, however. When wet or covered with manure, it can become quite slick. Plywood is an especially poor choice for flooring beneath hogs of nearly any age. Treated wood can sometimes burn sow udders and the delicate skin of very young pigs; white-skinned

This is a humane farrowing crate with a bottom bar that turns for pig safety and sow comfort. Notice the added space behind the sow and in the pig bunk area. Many modern crate designs protect pigs early in life while still giving the sow enough room and comfort.

NEWER CRATE OPTIONS

Among the most recent modifications to the basic farrowing crate are steel fingers and a hydraulic action on the bottom crate member along each side of the sow. As the sow begins to settle down, she is discouraged from dropping down on any pigs in her zone, and the rotating fingers push the pigs away from the sow. Another modification even uses puffs of air to move pigs away from a descending sow. Such state-of-the-art features add substantially to the costs of a farrowing crate, so many of them are best suited for the newer versions of the environmentally controlled farrowing units.

pigs seem especially vulnerable to such irritation. Also, very rough flooring sometimes causes a problem with navel ill. This is an infection of the umbilical cord stump that you'll learn more about in chapter 10, but it can begin after baby pigs' delicate underlines are irritated by lying upon or squirming across rough wooden surfaces. Navel ill can sometimes even result in umbilical hernias; little gilts can suffer irreversible damage to their teats.

I once had a problem with navel ill that started when pigs crawled over a 2×4 nailed to the pen floor to support a pig-bunk partition. We removed the offending board, and the problem was quickly eliminated, but we remain sensitive to what happens in the little pigs' environment, which is so easy to overlook when going about chores and seeing and experiencing things only at head and shoulder level.

Over time, I have seen folks fasten rubber pads to the floors beneath sows and pigs, as well as try combinations of slats and solid flooring, various types of expanded metal and mesh floorings, and even combinations of wood and mesh. In vogue right now is the practice of elevating the whole crate 8 to 10 inches above existing solid floors, placing 5 × 7-foot wire-mesh pads beneath the crates, and letting wastes fall into shallow pits also set beneath the crates.

Normally, sows go into farrowing crates three to four days ahead of their due date to become accustomed to and at ease in their new surroundings. Following farrowing, they may remain in the crates anywhere from a few days to until the pigs are weaned, at between 14 and 56 days of age.

Following are some additional tips for management of animals using farrowing crates:

- The sows should be treated for both internal and external parasites shortly before being placed in the crates. Some producers give the females a good scrubbing down with mild soap and water first, and some have even devised special "sow showers" for use before placing them in the farrowing crates.
- The sows can be fed and watered in the crates, but the equipment to do so properly can be rather costly. Also, anytime water goes into the crate, problems with dampness and slick spots in the floor can occur.

- In many smaller herds the crated sows are turned out twice a day, night and morning, for about 30 minutes, to eat, drink, and stretch their legs. Many sows will even wait until they are out of the crates to dung and urinate. With this system the sows are generally kept more comfortable, and while they are outside, you can care for pigs and clean or rebed the crates.
- If the weather is really hot, you may want to let the sows out of the crate more than twice a day. You can also use fans and vents to keep them comfortable.
- In cold weather, I always lay sheet tin directly atop the farrowing crates and use straw bedding to help hold in body heat.

Farrowing Stalls

Farrowing stalls are an old, old farrowing option that still has a lot of adherents. Such stalls are generally 4 to 5 feet wide and 10 to 14 feet long. The back 2 to 3 feet of each pen is partitioned off for baby-pig bunks. We fitted six such 5 × 12-foot stalls into one bay of our old Missouri horse barn, sealed it up with plywood sheets on hinges that could be raised to increase air circulation in hot weather, and farrowed in it quite comfortably 12 months of the year.

Stalls give animals a great deal of freedom of movement and can contain both sow and litter easily until weaning. Pig losses to overlay will be a bit greater — although 2×4s attached on edge down each side of the pen will do a

FARROWING CRATES AND HUMANE PRACTICES

Yes, the farrowing crate has been much cussed of late, but let me say first, we're not talking about a gestation crate that is used to confine a sow for months on end.

The crate's role on the small farms of the Midwest has been to contain sows for a few days ahead of farrowing and then for a week or two after farrowing. And for an hour or so twice each day, the sows are turned out of the crates for exercise, dunging, food, and water.

One to two weeks after farrowing, sows and litters are moved to pastures or lots. The move is based on the weather and the age of the pigs. They are turned out as soon as they can follow the sow and adjust for the weather. In extreme seasons, they may be kept inside for a few additional days, but the sooner they hit the dirt the better for most small-scale producers.

Crates do measurably increase baby-pig survival, make the sow and litter easier to access for health care, enable supplemental feeding on an individual basis for smaller and younger families, and can get baby pigs off to a better start. Over the years, we have used 5 × 10-foot farrowing stalls, farrowing crates, and one-sow houses, and all have strengths and weaknesses.

A neighbor is enrolled in a "humane" swine program that enables him to use a modified form of farrowing crate for the first few days of the pig's life. After three to five days, the crate sides are made to lift or swing away, and the crate becomes a stall with sheltered pig bunk.

The farrowing crate continues to be useful on many farms. It has been abused on some large farms where crating 'em up is used simply to pack in more numbers. Ultimately, it is the stock raiser that assures humane production.

surprisingly good job of protecting very young pigs from being crushed as the sows lie down to nurse. The 2×4s force the sow to lie down more slowly and away from walls. Pigs also may be sheltered under fence guardrails.

If allowed too much straw or other bedding material, a sow might build a large, mound-type bed or nest on which to farrow. If she farrows while lying atop such a mound, her newborn pigs may not be able to reach her side to nurse. These pens require extra labor to keep them clean, dry, and safe.

The Pull-Together

A third option for a farrowing enclosure entails incorporating crates or stalls into a modular-type building called a pull-together. This is normally built in two halves, with a length for each half of 10 to 30 feet, a width of 7 to 8 feet, and an additional roof overhang of 3 to 4 feet; the two halves can then be pulled together to form a modular farrowing building with a large central alleyway.

Depending on its size, the house may contain four crates or more or stall up to 10 or 12 sows. The larger halves may require a tractor with at least 50 horsepower to tow them about and position them. This is, however, a housing option with many virtues. Consider the following:

- Per crate or stall, the pull-together is one of the least costly of all farrowing options. Our first 8-crate house cost just $650, and a 10-crate house can now be bought new for less than $3,000. A used house can be found for just about a third of that.
- The house can be taken apart easily, to be used throughout a pasture rotation

system and as a further sanitation measure.

- It is of a very handy size but does not appear on property-tax rolls as a permanent structure.
- For income-tax purposes, it will often qualify as a depreciable, special-purpose expense. Its cost may even be entirely deductible in the year of purchase in many circumstances.
- It can be moved from farm to farm should you ever decide to relocate.
- It is suitable for farrowing 12 months out of the year.
- With only minor reworking, it can be made usable for other species; for example, for lambing, rearing bucket calves, or even housing poultry.

Single-Sow Units

The option I favor here at Willow Valley is the single-sow-and-litter farrowing unit. In many parts of the country, the solid-floor version of this 6 × 8-foot house is called a "Smidley," a trade name taken from the Ohio firm that makes the distinctive orange-colored farrowing units, as well as other livestock equipment. In this same vein, there are simpler-still three-sided, floorless huts that are used for seasonal pasture farrowing.

My choice for year-round farrowing is a row of 6 × 8-foot farrowing houses fronted by slatted-floor pens of the same dimensions as the house and then set 8 feet apart. That 8-foot spacing is adequate to prevent the aerial spread of disease organisms.

Either the houses can be set up as 6 × 8-foot self-contained farrowing stalls, or farrowing crates can be built into them with 2-inch planking. With either option 2 × 4-inch guardrails should be installed down each side-wall 8 inches above the floor.

I favor a house with a walk-in door on each end. The doors provide good access to the sow and litter, especially through the rear door into the pig bunk. Roofs can either slide back or lift up, which is useful in regulating airflow across the animals in hot weather and makes cleaning easier.

SEASONAL FARROWING ON THE RANGE

Seasonal farrowing on range is possible six to eight months of the year in most parts of the country and even is pursued in Michigan and Minnesota. On pasture, the houses need to be positioned 100 to 150 feet apart, to keep sows from doubling up in them and thus increasing pig losses through crushing.

The doors are generally 18 to 20 inches wide and 30 to 36 inches high. A piece of a 34-inch-high hog panel cut just a bit wider than the rear door opening can be fitted across it in warm weather; then you can leave the rear door open to further improve airflow. And in winter here in east-central Missouri, with the addition of just one or two 125-watt heat lamps, these houses do an excellent job of sheltering a sow and litter. In a tightly constructed house with both doors shut, a sow and her litter will generate as much as 6,000 BTUs of moist heat per hour.

Most three-sided huts are 4½ or 5 feet wide by 7 feet long. They are generally made entirely of sheet metal on a lightweight metal frame and resemble small Quonset huts or are of corrugated sheet metal on a wooden frame. Such houses are often so light that one person can load and unload several of them from a pickup — even flip them up and drag them about for positioning in lots or on pasture. If

This is a simple, one-sow farrowing hut.

not adequately staked down, however, some of the lower-height models can actually be moved around by big sows in a turtle-shell manner. Seeing a sow walking across the pasture wearing her house will certainly bring you up from the supper table with a bit of a start.

The sows need to go into the houses early enough for each sow to select her own house and stake out her territory at least three to four days before the first sow is due to farrow.

Housing for Sows with Nursing Pigs

Some producers set their houses in place first and then drop off a sow at each house site. The huts for range farrowing need some sort of bar or stop across the doorways to prevent the little pigs from leaving the safety of the huts until they are at least several days old and can navigate the nearby terrain and safely return to their own beds.

Long rollers can be attached across the bottom of the doorways for this purpose; a newer innovation is a small enclosure 8 to 10 inches high that encompasses a few square feet in front of the hut. Pigs can venture out a bit while still being safely contained, and the sows can simply step over the low enclosures to venture farther afield.

In very cold weather, the Smidley-type houses described under Single-Sow Units (see page 173) can be drawn inside larger structures such as barns or machine sheds. The animals are protected from strong winds and intense cold, electricity may be available for supplemental heating, and the houses can be drawn closer to home and feed and water supplies.

Traditional one-sow farrowing house

SINGLE-SOW FARROWING HOUSE

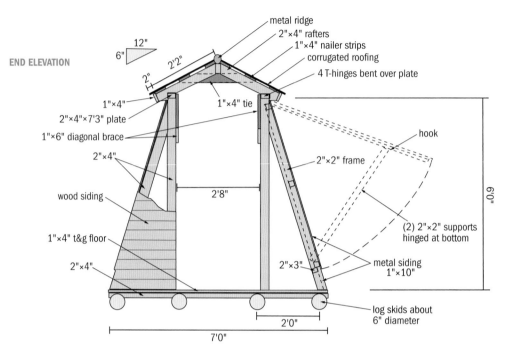

END ELEVATION

metal ridge
2"×4" rafters
1"×4" nailer strips
corrugated roofing
4 T-hinges bent over plate

12"
6"
2'2"
2"

1"×4"
2"×4"×7'3" plate
1"×6" diagonal brace
2"×4"
wood siding
1"×4" t&g floor
2"×4"

1"×4" tie
2'8"

2"×2" frame
hook

(2) 2"×2" supports
hinged at bottom

2"×3"
metal siding
1"×10"

log skids about
6" diameter

6'0"

2'0"
7'0"

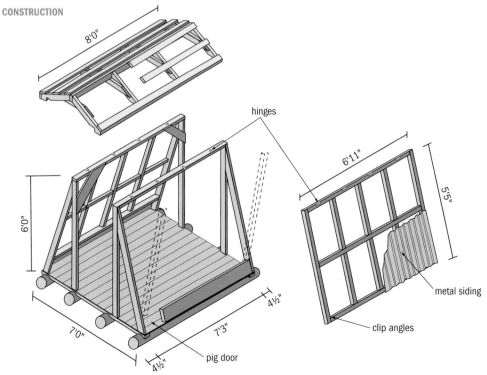

CONSTRUCTION

8'0"

hinges

6'0"
7'0"
7'3"
4½"
4½"
pig door

6'11"
5'5"
4½"
metal siding
clip angles

CREATING A COMFORT ZONE FOR BABY PIGS

We rely greatly on heat lamps and pig bunks at Willow Valley, but there are hog producers who use all kinds of combinations. Many of them "tinker" until they find the combination that works best for them. Here are some of your options:

Hovers are three-sided boxlike enclosures that fit in and over a portion of the pig bunk and are accessible only to the pigs through small "pop" holes, creating a snug, easy-to-heat zone. Made of plywood, they simply pull in and out of the pig bunk. Some fit over a heat pad; with others an electric heat lamp shines down through a circular hole cut in the top.

Heat pads are made of a sort of armored plate or fiber, attached securely to the floor, and positioned well away from the sow. They are regulated with a thermostat, and the pigs are warmed by lying atop them. For a few days after farrowing, you may have to use a heat lamp to draw the little pigs to the area warmed with the mat.

Red-tinted heat bulbs seem to do a better job of drawing the pigs to hovers and bunks than clear or white-frosted bulbs.

Insulating material can be taken right down to floor level in farrowing quarters if it is protected from damage by the sow and pigs. Covering the insulation with 1 × 2-inch wire mesh will shield it from the hogs.

An 8 × 10-foot pickup tarp drawn tightly around the end of the house that holds the pig bunk will greatly reduce drafts in cold weather. A few bales of straw across the bunk and up the house sides will further insulate the house. One very cold and snowy Missouri winter, I recall seeing neighbors heaping snow around and over houses in somewhat of an igloo fashion.

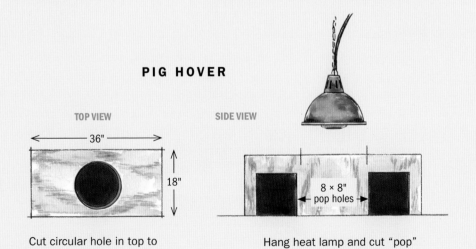

PIG HOVER

TOP VIEW   SIDE VIEW

←——— 36" ———→

18"

8 × 8"
← pop holes →

Cut circular hole in top to
let in heat lamp warmth.

Hang heat lamp and cut "pop"
holes for little pig entry.

Readying Supplemental Heat Sources

You should have a supplemental heat source for young pigs ready to go before farrowing, as sows can be bred to farrow 12 months out of the year. There are a number of ways to provide the heat very young pigs need to stay safe and warm. The trick is to warm the pigs without causing heat stress to their mother.

Among the heating options are forced-air units with the capacity for heating whole buildings, flameless gas heaters that use ceramic heating heads, electric heating mats that warm the pigs from below, and electric heat lamps. Wood heat is also often used.

One Amish hog raiser of our acquaintance uses wood heat to warm his farrowing house, supplied in a novel way. The firebox sits outside the farrowing house on one of its ends, with the flue on the other. Heat and smoke are drawn through a large pipe in the farrowing-house floor. The heat radiates upward through the concrete floor to warm the young pigs at the back of the farrowing stalls. It provides a fairly efficient comfort zone for the young animals.

Trying to heat a whole building to the temperature at which very young pigs are most comfortable can be quite expensive. It can also cause heat stress to the sows, with resultant restless behavior and a decline in milk production. A sow is most comfortable at 55 to 60°F (13 to 16°C), whereas a newborn pig has just left an environment with a 90°F (32°C)-plus temperature. The best compromise is to create a zone in which the pigs can be kept comfortable without overtaxing the sow or the pocketbook.

SAFE USE OF ELECTRIC HEAT LAMPS

On most small farms, the primary utensil for providing supplemental heat in any season is an electric heat lamp with a metal reflector. It is lightweight, inexpensive, and simple to use — and can burn down a barn in a New York minute if not managed properly.

Safety guidelines for use of electric heat lamps include the following:

- Never use more than seven of the 250-watt heat bulbs on a single circuit. The 125-watt bulbs will do a good job in many months of the year and will cost measurably less to operate.

- Never suspend the lamps by their electric cords or fasten them up with string or baling twine. You can use smooth wire to suspend the lamps, but the safest choice is lightweight chain.

- Never suspend the lamps lower than 24 inches above the pigs. The lamps warm by heating what they shine on and thus can badly burn tender-skinned young pigs or even ignite bedding.

- Be sure to keep raising the lamps as the pigs grow. A pig, which can rear up to a rather surprising height when even a few days old, can pull down lights or cause bulbs to burst by touching a burning bulb with the tip of its nose.

- The lamps must also be suspended in a position well away from the reach of the sow. Sows will rear up and climb on gates or crate rails to reach heat lamps.

- Steadily raise the lamps to wean the pigs from their dependency on the extra heat. Seldom do I use them for more than 14 days (generally 7 to 10 days). We often replace 250-watt bulbs with 125-watt within 72 hours of farrowing.

- Check the lamps often while they are in use for frayed cords, dust buildup on the reflectors, any slippage out of position, and loose bulbs.

- In extreme weather, we have used two lamps to get a young litter through a cold night. In such conditions, a second lamp suspended above the sow's hindquarters at farrowing can increase pig survival. It must be monitored quite closely, however.

- If outside air temperatures are above 60°F (16°C), you may not need supplemental heat.

8

SELECTING BREEDING STOCK

To select breeding stock, you build on the principles of animal selection outlined in chapter 2, but you must also factor in the type traits and genotype that influence reproductive performance. Of all of the economic and carcass traits you will hear about when discussing swine selection, none is more important than live young. To get these, you have to have animals with the will, vigor, and conformation to do those most basic of tasks: live and reproduce.

Genotype is the two-dollar word for the inward genetic makeup of an animal. It is what the animal can do for you in such areas as growth rate and carcass type. **Phenotype** is how the animal's genetic makeup manifests itself visually. It covers traits such as conformation, soundness, and muscling that are at least somewhat apparent to the naked eye.

There is no more important or better place to spend money than on the acquisition of quality breeding stock. In breeding stock you do indeed get what you pay for — but if you shop carefully and invest wisely, you actually get a bit more.

Shopping for Livestock

Several years back, the pork market was going through one of its boom periods, when anything female and 250 pounds or heavier sold as if it were gold plated. As a firm believer that the time to start or add to a livestock venture is when the desire and the dollars come together, I was in the market for a couple of extra gilts.

Down at the sale barn, gilts were busting butcher-hog prices with $100 to spare, the seedstock companies were making car dealers look like pikers in their advertising, and breeder auctions were followed immediately by trips to town to see if Rolls-Royce made pickups. Yet in just 20 minutes on the phone, I found a set of gilts less than 30 miles from home that were priced only $75 per head over market, had good pedigrees, and were guaranteed to breed and settle.

I've found the best ways to find out about good buys on hogs are (in order of preference) talking with fellow raisers, attending hog events, reading local newspapers, and checking out hog publications.

This old girl epitomizes the adage "as happy as a pig in the spring sunshine."

To eliminate the disease **pseudorabies**, which is spread by nose-to-nose contact, far fewer sale barns and community auctions now sell breeding animals, especially not breeding boars. I think this is a good thing, because a lot of folks were buying boars there that weren't paying off. Most boars today are purchased from private breeders.

Swine production is one of the few ventures in which you can get on the phone and order up hogs sight unseen, knowing that they are backed by a code of fair play adopted and sanctioned all across the industry. This is a semiofficial code adopted by a number of breed associations, with the basic element, of course, that the hog can be returned if the buyer is not satisfied. It sure does make seedstock shopping a whole big bunch easier. Still, there are a few things that the savvy shopper does need to do.

Compile a list of prospective suppliers. Contact them first by phone, and make appointments to view the hogs if they sound promising. It's just not fair to roll into a driveway and expect a producer to turn off a tractor at the height of planting or harvest to show you hogs.

Start your seedstock shopping well in advance of need. Going to look for a boar when the gilts back home are already going through their third heat cycle marks you as a real rube. It can also get you saddled with a sale-barn sow settler — an animal that just won't advance the herd — rather than a boar that will move your venture forward.

Select seedstock with the hope of improving or upgrading no more than one or two traits at a time. Try to key on the most glaring fault in your sows or their offspring and work at improving it.

Buy from facilities as similar to your own as possible. This helps minimize stress on the animals.

Try to view siblings, sire, and dam on the seller's farm to ascertain just how dependable the genetics really are. Are they truly breeding on and breeding "deep" — that is, are they producing quality animals in good numbers?

Don't hesitate to ask questions on the seller's farm on everything from how the hogs were bred to how they were fed. A good animal might be redeemed from neglect or poor nutrition but not if it has been stunted, nor will the best of care overcome poor genetics.

Don't haggle over price. If it's too high, thank the producer kindly for his or her time, and look on down the road. That way, you can both feel comfortable about possibly doing business together in the future. And be realistic: If your budget is tight, don't go looking at state fair winners.

Form a mental image of the kind of hogs you want, and carry it with you into the lots and pastures that you visit. Be fair, however, and realize that the perfect hog has never been bred. It is perhaps best to go with first impressions when selecting breeding animals.

COLOR CONSIDERATIONS

An old rule of thumb is to buy colored pigs for cold or damp weather, because they will be hardier due to their colored-breed parents, and to buy white pigs for hot weather, because they tend to have greater heat tolerance.

Performance Data

When you're shopping for breeding stock today, you can encounter a deluge of statistical data. Through various scanning and probing devices, many of them ultrasonic, it is now possible to gauge loineye area (in square inches, the measure of lean meat contained within the loin), fat cover, and muscle mass on a living animal without harming it in any way. Actual slaughter data may be available on part and full siblings. Performance data may also be available on one or both of the animal's parents, and on some proven breeders there is sufficient collected data to estimate a bloodline's influence on a number of specific traits.

Many years ago, when the trend toward collecting performance data began, a set of minimums for boar performance was established and almost codified. A boar being considered for service had to have at least a 4.5-square-inch loineye and a maximum of 1.5 inches of backfat depth, and he had to reach a marketable weight of 220 pounds at 160 days of age. There have been many changes since this early set of standards, though, and nothing quite so constant is now in place.

Days are now counted up to 230 pounds or even heavier, backfat thicknesses have slipped to 1 inch and often much less, loineyes on boars now have 5.5 square inches as a minimum (6-plus square inches is now quite common), and butcher hogs regularly top 70 percent lean yield. Hogs tested for growth and feed efficiency (such tests are conducted by individuals, universities, and breed groups) were once producing 1 pound of gain on 3.5 to 4.5 pounds of feed. Now that figure is closer to 3.5 pounds of feed, including the feed consumed by the hog's sire and dam during breeding and gestation. Tested boars have begun closing in on the 2:1 feed efficiency ratio once thought possible only with broilers.

Indexing Pitfalls

Performance testing refers to individual data collected on each animal for each specific trait and is the most reliable method of assessing that animal. Beware of data presented in any other way — at least to an extent. For example, information is sometimes presented in the form of indexes that give a numerical score, over or under the herd average, for performance of a particular trait. The average is often given a numerical value of 100; better performers will score above 100, and below-average performers will receive a score

MINIMUM BOAR-PERFORMANCE STANDARDS

The minimum performance standards for boars have changed considerably over the years. Following are old standards compared with the widely accepted standards of today.

Performance Factor	Old Standard	Today's Standard
Days to market	160	160–175 (leaner hogs grow slower)
Market weight	220 lbs	240–260 lbs
Loineye	5 sq in minimum	6+ sq in minimum
Backfat	1.5 in maximum	1 in maximum
Lean yield	60%	70% or more
Feed:gain ratio	4.5:1	3.0–3.5:1

MISLEADING FIGURES

I can recall a noted Duroc boar of a few years back that quite literally tore up a major Midwestern boar test and rewrote the record books. His offspring, however, failed to perform up to expectations, were lacking in desirable conformation traits and good breed character, and often had soundness problems.

At one of the public displays of this boar at a breeder's auction, a veteran farmer sitting next to me in the stands summed up the animal and his value quite succinctly. He said, "The day they went to buy that boar, they should have spent more time looking at him and less time reading about his figures in the sale catalog."

below. Seldom will scores move more than 5 to 10 points above or below that average figure.

I find indexes to be a bit deceptive. A boar scoring a 98 for growth is measurably below the herd average, but many of us with a public school education remember 98 as being a quite good score. A score above 100 denotes performance above the herd average, but what if the average wasn't all that good?

The recent trend toward exceptionally lean hogs and the corresponding slowing in growth rates had a lot of sellers opting for indexes to score growth traits. Some exceptionally lean hogs may take as many as 10 to 20 extra days to reach good market weight.

Changes in statistical goals and uses have not come without other consequences. With the leaner hogs not only are growth curves slowed, but traits such as litter size

and reproductive performance can also be demonstrated to have declined. Further, there are several very important traits that cannot be denoted statistically.

Weighing the Data

Performance data is an important selection tool, and while estimated breeding values or performance differences are still very new and much debated, they are dramatically building the impressive numbers that can now be presented with some hogs. All of these numbers need to be weighed very, very carefully, however, as they are not practically obtained.

The way data is obtained and released is still very much under the control of individual breeders, breed associations, and test stations. The performance of young boars is enhanced by the presence of male hormones that will be unavailable to barrow or gilt offspring. And some data is obtained in ways that are neither realistic nor practical for most family farms. Test groups are often quite small and are often full siblings; they're quite closely confined and often fed rich, rather costly and complex rations to maximize performance. They are burning high octane to develop data that is supposed to be used in a real world run on regular gas.

Keep in mind that the traits for which most of the data is available are the traits with the highest degrees of genetic inheritability. Those with a lesser degree of genetic inheritability are no less important to the overall success of a swine operation, however. Litter size, for example, has a quite low degree of genetic inheritability (it is a trait as much influenced by nutrition and environment as by genetic factors), but if we let selection for litter size slide in our Duroc herd, it soon hits us in the pocketbook and hurts our boar sales.

We'll get into more particulars later, but other factors that must be weighed during seedstock selection include these:

- Litter size at both farrowing and weaning
- Length and body capacity
- Foot and leg structure
- The development of reproductive organs
- Muscle pattern
- Frame size
- Sexual character

To determine an animal's strengths and weaknesses in the above areas, you have to work the hogs (move through them), watch them move about, view their home environment, and learn something of their history.

I began my hog-raising career with a sow about halfway through her productive life, but it is far more common to begin with open or bred gilts. **Open** is a producer's term for a female of breeding age that isn't bred. Some gilts also are sold as exposed or pasture exposed to a boar, but there's no assurance that they are pregnant. Gilts will always be a bit of an unproved commodity, but a long, productive life should lie before them. It is easier to determine if a young female has merit as a producer of meat animals than it is to tell if a sow has this potential; pregnancy and nursing alter a female's appearance, and you won't find the tightness and definition of muscling in a sow that you will in a young gilt.

CULLING

Long ago we learned a lesson in stock raising that still gets ignored on some farms. You can raise a more vigorous and healthy herd through breeding practices.

Simply by selecting herd replacements from the largest and most vigorous offspring in each generation, you are naturally selecting for hardiness and the will to thrive. Match this with culling for soundness; deep, wide chests; body capacity; mothering; thriftiness and durability; and you are creating animals with the physicality to live, grow, and otherwise perform even in the presence of some harmful organisms.

It takes time, and you may have to cull rigorously in the early generations, but you will create a herd that will need far fewer injections and antibiotics in the feed rations. This is especially true when herds are kept in modest numbers and in an open environment.

Nature ensures that her herds will survive and flourish by removing the smallest, weakest, and most impaired within the herds. When producers use these same practices, only the fittest survive and reproduce in kind. The best herd for the small farm or smallholding will be deep in rugged, solidly made animals with a long history of selection for simple vigor and durability.

For the last decade that we were in heavy production on our farm, we did little more than treat for parasites and give iron shots to baby pigs. Many a bottle of antiseptic met its expiration date on the shelf. And some gilts and sows were not allowed to enter the breeding herd because they did not hold together as we knew they should.

Gilt Selection

The function of the gilt is to bear and mother the young, but this does not mean that you can neglect growth and meat type during the gilt-selection process. At one time, it was a fairly common practice to pull replacement gilts from the last hogs remaining in the finishing pen. This was actually selection for slow growth and weak type.

Home-Raised Gilts

You should select the biggest, most well-grown and -developed gilts in the group, even if it delays a payday for a week or two. If you are building a herd with home-raised gilts, initial selection should come from the best half of the early pig crops. In time the quest for quality should surpass the need for numbers, and only the top 10 percent of the gilts produced will be good candidates to enter your breeding herd.

In a well-managed herd, sows should be regularly producing gilts that are better than they are, which should further facilitate gilt selection. As an average, most swine producers cull and replace with new animals something on the order of 40 percent of their sow herd yearly.

In the selection of home-raised gilts, the first cut is generally made on the day the pigs are born. Gilts can come from litters farrowed by first-time gilts or by sows that are old hands. Keeper gilts should only come from gilt litters in which at least eight large pigs, uniform in type, were farrowed. Select gilts from sow litters of at least nine pigs.

Some hog producers balk at selecting breeding animals from gilt litters. The first reason for this is that the gilts have no proven track record as to reproductive performance. Also, due to their smaller size, gilts are more likely to have been bred to younger, equally unproven boars. Another factor to consider is that older sows have had a longer time on the farm and thus a longer exposure to the organism population, or "bug soup," that is to be found on all livestock acreage. Through this exposure they have built up a stronger natural immunity to those organisms — one that they pass on to their offspring.

WHY HOME-RAISED GILTS?

I am a firm believer in the home-raised gilt. A gilt's lineage, her temperament, and even the quirks of her dam and granddam should be fully known to the producer, and when she is raised at home, there should be no question about her history. You can grow and develop young females specifically for a long and productive life in home-farm breeding pens. Simply put, sow-herd quality control can begin the day the gilt pigs are born.

Economics still favor the home-raised gilt for small and medium-sized producers. Assembling a group of 5 to 10 gilts from outside sources good enough to replace a breeding group in a well-managed small operation is neither simple nor inexpensive. The goal with herd replacements is not to equal but to improve performance and to ferret out and then select the cream of the cream of the crop, which can be time consuming and costly if you're purchasing the replacements from elsewhere. But if you're doing your job right, you should find plenty of good ones inside your own fences.

The Gloucester Old Spots are of a somewhat older swine type, but this gilt has good bone and a better topline.

This sow has done a good job. She has a litter of eight pigs that are uniform in size, well grown, and of good breed character. Note the good underline on her.

Select Larger Pigs

It is a simple matter to take very young pigs in hand and give them a thorough examination. The best candidates for survival and long, productive lives are pigs that weigh 3 pounds or more at birth. I make special note of gilt pigs with a birth weight of 4 pounds or more; we have even had some pigs crowd 6 pounds at birth. By always selecting for larger gilt size over a period of time, you will emphasize the natural growth and female capacity that should also ultimately result in fewer farrowing problems and larger, more vigorous pigs at birth.

Underlines

Very early in gilts' lives is also a good time to check out their **underlines**. To be considered as a female herd replacement, a gilt should have at least 12 evenly spaced and well-formed teats. Fourteen is even better, and uneven counts of 13 or 15 are not to be discounted, although there should be a minimum of 6 teats on each side of the underline. Do not select gilts with unevenly spaced, inverted, or pin (too small) nipples.

Determine Growth Rate

The next cut in a home-raised gilt's life comes at about 150 pounds. First determinations about growth rate and efficiency of performance can be made then. Some producers will remove gilts to a separate developmental group from the finishing group at this point, but I prefer to leave them on a good 15 percent crude-protein growing ration and keep them with their peer group, to see if they continue to perform up to expectations.

The goal is to have a gilt weighing 300 pounds by 8 months of age, the age at which she should be bred if she is to farrow at about 1 year. With experience, I came to prefer breeding gilts at 10 months of age and weighing 300 pounds or more. At 220 to 250 pounds, keeper gilts should be sorted out and grouped separately, pulled off the self-feeder, and then developed steadily to that 300-pound goal. At each cut, evaluate the gilts for soundness, continued rapid and efficient growth, continued feminine character (see page 190), and meat type. Even at 300 pounds, gilts culled from the keeper pool will retain good value as butcher stock.

Backfat

At 5 to 6 months of age and 220 to 250 pounds, the gilts can be evaluated with one of the many scanning machines and probed for backfat. Most land grant colleges offer these services through their Extension offices, and a call to the local county agent should produce a phone number for these tests.

Uniform udder segments with well-formed nipples

Money in the bank: a set of Berkshire gilts of good quality and showing the uniformity that small farmers need to bring to livestock production to hold a position in the marketplace

BACKFAT PROBING

You can learn to do backfat probing yourself. It requires only a good catch crate to securely hold the animal in position, a scalpel, and a thin metal ruler with a sliding gauge. Standard procedure is to make three small incisions along the topline, probe down with the ruler to determine thickness, and average the three measurements to develop an average backfat thickness, a good guide to the animal's overall fat cover. A simple procedure that will work for those selecting gilts for home use is to probe just once in the third position, farthest back on the hog, and add 0.1 inch to that figure. It results in an approximation but is still a good, usable figure.

These days, it is generally possible to adjust performance data backward or forward to a standard of, say, 230 pounds. This allows you to compare the performance of a number of hogs of different weights and ages to a true constant.

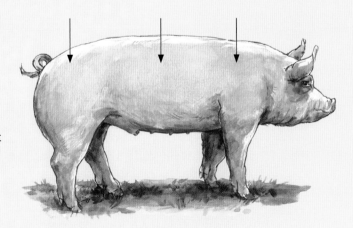

The three sites you can probe to determine backfat

Conformation and Appearance

A good gilt will have a wide, deep body reflecting a capacity not just to grow but also to carry a large litter of pigs safely to term. Currently in the United States, most females don't make it past their fourth parity, but on small farms sows tend to hang in longer; they have what it takes to endure.

Long, sloping pasterns, a well-formed leg set correctly on each corner, evenly sized and shaped toes, and a long and level top all contribute to long-term, continuing soundness. Large, well-formed bones as indicated by leg diameter are further indications of continuing soundness.

It is almost too abstract a term to explain, but a good gilt does have a truly feminine appearance. She must at once be large and well grown yet retain elements of refinement and smoothness. Overall, a gilt will lack the stature and projection of masculinity of a male, will show refinement in and around her head, and will project a feminine image. Since our bread and butter is breeding boars, from time to time I have tried to use a large, coarse-headed female to create more masculine-appearing boars. Sometimes it works, but with well-formed, feminine-appearing gilts and sows, I get just as many keeper boars — or more — in more even numbers and litters.

There is no device that allows the prospective buyer to peer into a female hog and view her reproductive tract, to evaluate her ability to cycle, breed, and carry large litters of pigs. The one visual clue to the capacity of the reproductive tract is the **vulva**, the only external organ of the reproductive system. It should be large and well formed. Very small vulvas are the one visual indication that the rest of the reproductive system might not be adequate.

This Yorkshire gilt could use a bit more frame and overall substance.

BALANCE IS BEST

Gilts too extreme in type can sometimes wind up sacrificing performance in the breeding pen or farrowing unit. For example, gilts too trim along the bottom line and that fail to carry depth of side throughout the full length of the body may develop what is termed a "meat-type udder." Quite simply, the udder does not fill correctly, and one or two teat segments on each side of the udder cannot be used by the nursing pigs.

In selecting gilts, try to avoid such extremes in type. Go with what most call the middle-of-the-road type, which balances reproductive type with at least moderately good meat type.

Boar Selection

There is no faster way to bring about change or introduce new blood into a swine herd than through the purchase of a first-rate young boar. Something akin to heterosis (hybrid vigor) has been noted in even purebred herds when a new boar of separate and distinct breeding is introduced. It is also the least costly and least complicated way in which to bring new genetics to the small farm.

As I've explained, boars usually are bought from breeders, and bringing in a good one is a great way to improve your breeding herd. Actually, two good boars used in succession can effect as much as an 85 percent increase in the performance of a herd's progeny for a great many important traits.

It is also widely held that the real benefits of a good boar are not seen until his daughters enter the breeding herd and begin producing offspring of their own. This points up just how much you have to be concerned about when seeking out a good boar to head up even the smallest of herds. His influence, good or bad, will be felt for generations.

I have been to many breeder auctions at which the breeder, auctioneer, and breed association field men worked long hours to select the right boar with which to start the sale and, presumably, bring top dollar and so set the tone for the entire event — only to see that animal outsold by one appearing many places farther down in the sale order. Either the buyers don't agree with their assessment or the buyers are in the market for a boar with particular traits that the lead boar doesn't have. It all depends on who's shopping for what that day.

An old chestnut of wisdom in hog circles holds that the boars that make the big splash at sales or shows seldom produce sons and daughters that go on to make as big a splash. Their impact isn't generally felt until the second or even third generation, and a lot of big boars are never heard from again after their moment in the winners' circle.

Still, selecting a herd boar isn't like having a top-10 choice in the pro football draft. You don't go after the biggest gun available; you go for the boar that will solve a specific problem — your most pressing problem.

There is a two-fold task in selecting a herd boar: Bring in the genetic piece or two needed to strengthen the one or two breeding weaknesses most common in your sow herd or their offspring, and don't lose ground in any of the other traits.

As noted, most boars today are purchased from private breeders for $300 to $400. They usually come with some sort of warranty or guarantee of soundness and breeding performance. You'll read more about that under Boar Warranties on page 196.

TERMINAL BOARS

"Terminal boars" is a category of breeding boars that has only recently emerged. As I explained in chapter 2, a terminal hog is one used as a cross to maximize growth and muscling, since the pigs produced are intended to be sold for butcher stock. This term certainly points up the role a boar plays in imparting the economic traits — growth and carcass qualities — to the offspring of a mating.

Terminal boars are generally F_1 crosses of two of the more noted meat-type breeds — Duroc, Hampshire, Berkshire, Black Poland, and so on — that are used to impart an approximate double dip of carcass and growth traits to pig crops. An interesting cross along these lines is the Hampshire × Spotted. The Spot black breed brings with it a better litter size, although Spots are underutilized because they are not so easy to find.

Boar Age

Most boars are bought as untried youngsters at 8 to 10 months of age and weigh between 300 and 400 pounds. Some cash savings can result from buying a young boar 60 days or so before service age. This is not a bad idea, because any breeding hog you purchase should be kept on your farm and in isolation for at least 30 days before coming in contact with the breeding herd.

Maternal Background

As noted, a boar's greatest impact often comes when his daughters and granddaughters enter the breeding herd. You must therefore carefully weigh a potential herd boar's maternal background before plunking down hard cash for him, because it is those maternal genes that will be most directly expressed in the gilts he sires. Viewing the boar's dam and even granddam would be a valid part of any selection process, but on some farms this may not be possible because the breeding herd is kept closed and isolated as a health-care measure. After establishment, linebreeding is used, and new animals enter only after a long period of isolation and in very small numbers. In any event, when questioning the breeder, do not overlook the maternal performance behind any boar you are considering.

As a prospective boar buyer, you need to make inquiries as to the size of the boar's litter at farrowing and at weaning, how well the sow milked and mothered (this is often demonstrated by whole-litter and individual pig weights at 21 days of age and again at weaning), and how many litters that female has produced during her life in the herd.

SIZING UP GENETICS

When I go out to buy a new herd boar, I look for the biggest pig from a big litter and one that has at least two or three keeper-quality male siblings among his littermates. I consider this an indication of both the depth of the genetics behind the mating and the dam's ability to raise a quality product in good volume. Producers working with smaller numbers of animals must emphasize this level of production and consistency in order to stay competitive.

A Landrace with good frame. If bred to the gilt on page 190, he would produce some better F₁ gilts with plenty of mothering potential.

Soundness and Appearance

Soundness is an absolute must in boar selection, as his success as a breeder depends greatly upon his agility and athleticism. Hogs with hind legs tucked too far under the body, knocked knees, sickle hocks, or other foot- or leg-structure problems should be weeded out early in the selection process. A good boar walks out on a lot of bone and will remind you of a champion boxer striding confidently to the ring.

A bit harder to pin down on paper, but also important in boar selection, are ruggedness and masculine character. Boars, especially purebred pigs, change greatly as they mature and develop — there is often even a time or two in their developmental stages when they may actually appear to be falling apart. Purebred hogs have long been bred to grow and develop a certain way, and as a result, they have their gawky moments until they emerge as a "finished product." It is a widely held belief that the two best times to select a boar are on the day he's born, and when he reaches 300 pounds and begins taking on his mature characteristics.

You should certainly select for large, fast-growing boars, but there also are a number of seemingly small guideposts that veteran producers use to select boars with lots of bred-in growth potential. Among them are width between the eyes and foot or jaw size. Many believe that the young animal will grow to match or "reach" those features. I opt for the width-between-the-eyes measurement, because I believe it corresponds to body width and good internal dimension.

Not only do male hormones influence growth, but they also go on to produce a number of secondary sexual characteristics. Among those are a larger, coarser, and more masculine-appearing head; more guttural vocalizations; and a chopping activity with the mouth that sometimes produces large amounts of a white, foamy substance. The foam appears to be a natural sexual attractant to sows, and a young boar that doesn't do a lot of chopping and rattling gates in displays of sexual aggression is probably going to be a problem breeder.

It sounds almost too simple to need to be written down, but a boar should look like a boar.

A typey Duroc boar in the breeding pen being bred to white females for market hog production.

TAKE A TEAT COUNT ON BOARS, TOO

Checking the underline and making a teat count is important in boars, too. Teats are an important trait to emphasize in the selection of a breeding boar, because his stamp will be borne far more by his daughters than his sons. A minimum of 12 evenly spaced, large, and well-formed nipples should be found along his underline. Teat placement on boars is also a good visual indication of carcass length; I always try to select boars with at least three teats ahead of the penile sheath on each side of the underline.

Reproductive System

Unlike a gilt, most of a young boar's reproductive system is open for visual inspection. The testicles should be appropriately sized for age, not withered or misshapen, and carried in the scrotum in a slightly asymmetrical position; that is, one testicle should be positioned a bit higher in the scrotum than the other. Avoid a boar with testicles that are noticeably uneven in size. There is some statistical evidence that the larger the testes, the sooner the male will enter puberty, and the more fertile he will be. The **penile sheath** should be trim in appearance and fit closely to the underline. In a too-pendulous sheath, fluids can collect and create a pocket ripe for infection, which might damage sperm quality. Injuries to the penis and bleeding from the sheath or penis are both solid reasons for rejecting a boar.

This is an older boar in breeding condition. He's in his working clothes.

Tips for Buying Boars

Whether you plan to buy a boar at a breeder's sale or elsewhere, here are a few tips to help ensure a successful purchase:

- If you're going to a sale, request a catalog, and study it in some detail. Use it to compare the performances of the various bloodlines represented and to determine how consistently the line you're considering performs. It is a tool for the comparison analysis of available data and not simply a wish book.
- Call the producer ahead of time with any questions.
- Arrive early on the day of the sale to view in detail animals that you have marked in the catalog for consideration. This also gives you time for a short visit with the producer to discuss the offering.
- Take advantage of the breed association and auction company staff on hand. They are there to help you make evaluations.

- Remember that the order of the sale is generally based on merit, with the animals determined to be best selling first to help set price trends and the tone for the whole sale. Still, do not be surprised to see a littermate to the sale starter enter the ring much later and sell for a great deal less than his or her sibling.
- Have a walk around the whole barn and even a turn inside the sale ring. Sawdust or sand in the sale ring generally means hogs with a set of feet and legs to show off. A deep bed of straw in the sale ring can mean just the opposite.
- Look up all the siblings being offered to the hog or hogs you are considering. If there aren't any or they're way down in the offering, let this temper your thoughts, because a really good line should produce more than one quality animal.
- Go to the sale with a ceiling price for your bidding firmly in mind and stick to it.

Boar Warranties

There are a number of terms or conditions to boar guarantees, and they form a pretty good set of guidelines about the care and management of the young animal. Though warranties will vary somewhat, a boar is generally warranted under the following terms:

1. The young boar must be used in a hand-breeding situation for the first few matings. He must also be brought into contact with one female at a time, preferably a young sow or gilt close to him in size and in good, standing heat. Early trauma such as fighting with a larger female not in estrus can damage or destroy a young boar's breeding career. Many breeders reserve the right to test the breeding abilities of a contested boar, however.

2. The warranty is invalid if the boar is simply dumped into a pen-mating situation. We once sold a pair of 5-month-old boars that looked big enough to breed to the buyer, but we knew they really needed at least another 45 days of seasoning. When we delivered them, the buyer had us unload them right into a pen of gilts of several different ages. I'm afraid that buyer wasn't very happy when I told him those boars were strictly "as-is-where-is" — that there was no warranty if he violated basic rules of management and that he was subjecting them to use and risks far out of reason.

3. The warranty is invalid if any rings are placed in the young boar's nose. A boar often uses his nose to position the female for mounting and service; ringing can create boars that are shy breeders.

4. Many breeders require notice as soon as possible that problems are emerging with boars they've sold. Some limit the warranty period to 30 to 90 days following the date of sale. We have had less than half a dozen complaints in our nearly 30 years in the boar business, and two of the most vociferous were actually due to the buyer's neglect or slight. The most vocal was not about a boar's failure to perform but uneven litter size at farrowing. That was quite clearly a management problem, since uneven litter size is generally a nutritional problem, environment related, or the result of a health problem that develops following the relocation of the boar. For example, even low-grade fevers at the time of breeding can adversely affect sperm counts in boars and embryo survival in bred females.

5. The replacement clause in such a guarantee normally provides for the replacement of the boar with another of equal value, or a discount on the purchase of another boar.

Bidding Decisions

There is one final question regarding breeding-stock selection. What is a good one worth?

One of the founders of the legendary Wye Plantation herd of Black Angus cattle once stated that you should expect to spend at least the monetary value of your five best females to acquire a male that will truly advance the depth and quality of your herd or flock. Such a high expenditure may not always be necessary to maintain a high-quality swine herd, but it does point up the value that you must posit

An older Wessex boar. He may be getting too large and cumbersome for extensive use, especially with gilts. Breeding boxes were once used to accommodate older and larger boars of considerable value, but as boars approach this size and age, a younger boar should be brought into the breeding herd.

for additions to your breeding herd if quality is to be maintained. The time of widespread four-figure prices for swine breeding stock now seems to be well past. A true farmer/breeder now buys a whole lot of boar for $300 to $600.

Gilts are generally priced via a formula tied to current butcher-hog prices. Typically, this will have a 250- to 300-pound gilt selling for market top at a specific market point (often one of the Iowa markets) plus $25 to $100 per head. Bred gilts seem to fall fairly consistently within a $350- to $750-per-head range, with the top end for pedigreed females carrying pedigreed litters. The Missouri Pork Producers group regularly awards $1,000 to a number of 4-H and FFA youngsters, confident that these youths can each purchase two bred, pedigreed gilts at the state's annual spring-bred gilt sale.

Gilts bought in groups may be discounted a bit from the prices for gilts bought one or two at a time. Gate-cut gilts from finishing pens can sometimes be bought for a slight increase over what they would bring as butchers. And some good gilts can be bought at the various weanling pig sales.

BUY FROM FELLOW FARMERS

Want to keep small local farms alive? Then buy purebred seedstock, and buy it from your fellow family farmers.

The so-called seedstock companies spend a lot of money on advertising and provide a few ruffles and flourishes like 800 phone numbers and delivery services, but they, too, tap into the swine gene pool that is held and maintained by America's family farmers. A great many of the lines these companies offer are crosses and breed composites, and it is the companies that reap the greatest share of the hybrid-vigor benefits.

9

MANAGING YOUR BREEDING HERD

Once you select and purchase your animals, you cannot just take them home, drop them in a pen, and start counting off the days until the pigs arrive.

New breeding stock should be isolated for at least 30 days and during that time should be closely observed for any potential health problems. They need to be penned well apart from any other hogs on the farm and tended to last each time you do chores, to prevent you from tracking any potentially harmful organisms from their pen to other animals.

It's not practical for small farmers to maintain a completely closed herd, but we keep our breeding stock far from the animals for sale, on the opposite side of the farm. Visitors generally don't see the breeding herd, and during farrowing season we mostly stay home, because the health risks are greatest for little pigs. Chore clothes and footwear are not worn off the farm. We do not visit other hog herds or travel to sales or other events where farmers are apt to gather when we have farrowing sows and very young pigs on hand. And I very much discourage visitors in and around farrowing quarters and young pigs. More than one feed salesman has been run off a farm for entering a hog building unbidden or even driving onto the wrong part of a farm.

Bringing in the Boar

Especially critical is how a young boar is introduced into the herd, because even a slight injury or low-grade fever can knock him out of service for a month or more. Such a time lapse can badly skew breeding schedules and disrupt both farrowing plans and budgets.

There are a number of simple steps, however, that will help you ease a young boar into the breeding herd:

1. **Boost immunity.** After the quarantine period is safely past, many hog farmers transfer a few scoops of manure and old bedding from the sows' pen to the boar's pen, and vice versa to slowly expose them to various pathogenic organisms. Every farm has its own population of good "bugs" and bad, and this early exposure can prevent later health problems, especially fevers, which can cause temporary sterility in boars, or the failure of embryos to implant in a sow's uterus.

Setting newborn pigs to nurse

2. **Allow fence-line contact.** Pen the boar adjacent to the females so that they can sniff and have nose-to-nose contact with each other through the fence wire. They become more comfortable with each other this way, and the chances of fighting when the new boar enters the breeding pen are greatly reduced. Fence-line exposure can also cause a pen of females to begin cycling. These must be very stoutly made pens: 54-inch-high cattle panels set with posts on 8-foot centers are probably the best choice for their construction.

3. **One-on-one introduction during estrus.** The first few gilts or young sows in estrus should be exposed to the new boar one at a time and only when they are in standing heat.

4. **Observe.** You should be on hand to observe those first few services to detect any problems. Things to note especially include high riding, which can cause irritation and injury to the boar's penis. Look also for blood and ejaculate on the sow's back, on the penis when extended, or on the sheath. Blood kills semen. High riding generally is grounds for immediate culling, and an injured penis can be slow to heal. But some high riders do learn better technique through trial and error during hand mating.

Fence-line contact between the sows and boar is one way to ensure that the animals become more comfortable with each other and less likely to fight when the new boar enters the breeding pen.

The rule of thumb is to maintain one boar for every 10 sows on the farm. And even a young boar should be able to breed and **settle** (impregnate) half a dozen sows or gilts on a single cycle. Double-breeding the female at 12-hour intervals during her estrus cycle, or breeding her to two different boars, will also increase successful breeding percentages.

You've probably heard about artificial insemination (AI) in the swine industry (see page 134). Personally, I do not favor AI. Semen has to be used fresh, and the cost for purchase and transportation can be expensive. Also, the semen trade is in the hands of a few corporations now, and in my view, AI hurts boar sales for small producers.

Bringing in Gilts

Trying to bring one or two gilts into an existing group of older sows can be very trying; in fact, many producers choose to replace whole farrowing groups rather than attempt piecemeal changes. If timed to coincide with periods of higher sow prices, the salvage value of a heavy cull sow sold for butchering is often enough to buy her gilt replacement. If you retain and raise your own gilt replacements, keep and breed at least 10 percent more females than you will actually need, to be sure an adequate-size group will be in sync to farrow together.

New animals being introduced to a group should be closely watched for at least several hours. If a young female is too timid, she may never meld into a group, and her entire lifetime performance will suffer as a result. If you find yourself with such an animal, pen her separately, put some weight on her, and send her to town.

The safest and most useful feeding stalls can be sealed with a gate after the hogs enter.

INTEGRATING GILTS INTO A SMALL HERD

Gilts are younger, generally smaller, and more docile than sows, and if they are simply dumped into an established pecking order, there is a very real risk of serious injury to the young animals. To fill in one or two openings in a very small group of females, there are a few measures to try, although none is a guarantee of success. These include the following:

- Moving the sows to a pen where the gilts are already established

- Lightly misting all the animals with a strong-smelling liquid, such as kerosene, so that they all smell the same to each other

- Penning the animals in a very large enclosure with plenty of running room and extra sleeping sheds (this is our preferred option)

- Using feeding stalls

Be vigilant about watching new gilts at feeding time for several days, and consider using feeding stalls to confine animals individually. These stalls generally are 24 inches wide by 72 inches deep. They are made of native hardwood or treated lumber and are normally erected with 2-inch flooring on runners for easy portability. They can be made with or without a plank floor, but hogs usually tread the ground to mud beneath crates without floors.

The safest and most useful stalls can be sealed with a gate after the hogs enter them. Thus each female can eat safely, take all the time she needs to eat, and receive exactly the amount of feed she needs; also, younger and smaller females aren't dominated and harmed by the more aggressive "boss" sows in a pen. The stalls are also useful for containing females to receive individual health treatments.

Breeding Gilts

Gilts generally enter puberty at around 8 months of age (6 to 8 for boars) and cycle every 21 days thereafter until bred. To get gilts ready to breed, many use a practice called **flushing**. For 7 to 14 days prior to entering the breeding pen, the gilts are placed on full feed — roughly 3 percent of their body weight daily. For gilts that have been limit-fed to maintain condition or sows that have been dramatically nursed down, this gives their systems a sort of kick start.

As mentioned before, exposure to the boar can often start a set of females to cycling. Another old trick to get gilts cycling is to take them for a short haul in the pickup or trailer. Many farmers will pick a nice morning, load up the gilts, go to town for a cup of coffee, take the long way home, and then unload them into the breeding pen. This break in the usual routine seems to trigger responses in the gilts.

Other than watching for the boar's response, one way to tell if a gilt is in **estrus** is the back-pressure test. In the presence of the boar, put pressure on the gilt's back. If she's in estrus, she'll "stand." Her ears will come up, and her tail will twitch.

Just ahead of breeding is also a good time to treat the animals for internal and external parasites (check out chapter 10 for more information on this).

A few hog producers **hand breed**, which means placing each female in **standing heat** together with the male, witnessing the mating, and returning the female to a separate pen. They will then breed the females again in 12 or 24 hours. For this effort you'll have an exact breeding date, but the process entails added labor. In addition, sows penned away from boars may be slower to come into heat and have less pronounced estrus cycles. Some sows may even experience a **silent heat**, with no detectable signs of estrus.

What's Best for the Boar

One of the big challenges when working with small numbers of breeding swine is to maintain a breeding male in good breeding condition without too quickly becoming too large to be safely used.

A boar left constantly with a small number of sows will overeat at the expense of his penmates and can soon grow to a quite large size. A neighbor once had a boar less than 2 years old grow to more than 700 pounds because he fed with a scoop shovel throughout one long Missouri winter.

Hand mating is the ideal way to use a boar, but it is a time-consuming practice and one that will require added handling and facilities.

Over the years we adapted a form of pen mating that works pretty well on our small acreage. The steps we followed were:

1. Ahead of first breeding and after each weaning, the females were flushed. They would be placed on a full feeding of the gestation ration for 10 to 14 days prior to breeding. This was to have them gaining in weight and improving in condition as they were being bred.

2. The boar would be placed with them in a large lot with feeding stalls and remain with them for 30 days. This was to ensure a second exposure for any females that didn't settle or take during the first mating cycle. Its onset was generally triggered by the presence of the male.

    Estrus will occur every 21 days until the sow conceives. The estrus cycle will last for 72 hours and the best times for a successful mating to occur are 12 hours after standing estrus is first noted and again 12 to 24 hours later.

3. At the end of the 30-day breeding period, the boar would be removed to separate quarters. These need to be bull-horn stout in their construction, provide the boar with room to exercise, be well shaded, and be positioned out of hearing range and line of sight or smell of any females on the farm. Our boar pen was under a couple of red oaks behind the shop building.

4. Any females that do not settle after two opportunities for breeding should be considered for culling.

5. The boar penned apart will need careful monitoring to ensure that he does not become sour in his ways or gain too much weight that will then impact his ability to breed safely.

6. He should be given plenty of water, but fed only a maintenance ration. This will generally be met with 3 or 4 pounds of grain and a half pound of protein supplement fed daily to provide him with 0.5 to 0.6 pound of total crude protein daily. A bit more may be fed if he lost much condition during breeding or the sows are growing away from him. And he, too, might benefit from a light flushing before being put back into service.

   Especially with a boar, you should be careful not to feed a too finely ground ration as it can have an abrasive effect on his digestive system. To boars penned apart, we generally fed shelled corn or even corn still on the ear. A bushel of ear corn will weigh 70 pounds and should contain 70 good ears.

7. To provide some stimulus in his quarters, we might hang old tire chains from one of the trees shading the pen or put a couple of old bowling balls into the enclosure. I don't say that a hog will actually play, but they are creatures of some curiosity and such additions to the pen will keep the boars from pushing at fencing and housing.

8. A boar that has been out of service for some time may have early issues with stale ejaculate upon being returned to service. To counter this, a sow being readied for culling or a couple of market gilts might be penned with him a few days before returning him to the breeding lot.

   If that is not possible, observe his performance in the breeding pen to note that the females in estrus are being serviced multiple times. A young male may become fixated on a single female. Remove her for a brief time, and watch closely to be assured that he isn't being pushed back or fought away by an older or more aggressive female.

9. With our small herd, because female replacements were home-raised, we generally brought in a new young boar at 18-to-24-month intervals as new gilts were brought into production.

10. Too often a great deal goes into the selection of a boar and then he is treated like a big old lump after he is brought home. Lose a boar one time to injury or, worse still, death and the resulting scramble to keep the breeding program on line will make you much more appreciative of him and mindful of his care.

MATING SUCCESS

How do you tell if mating was successful? Following a completed service, a "semen plug" may appear in the vulva. It is formed from ejaculate material and prevents semen from leaking out of the reproductive tract. There may also be a rather pronounced flour-paste smell. Also look for dirt and hair disruption on the sow's back, which indicates that mounting occurred.

Gestation

A sow is pregnant for an average of 114 days —
3 months, 3 weeks, and 3 days, as the old
hands like to relate it. During that period, a
gilt needs to gain about 125 pounds to ensure
her continuing growth and good litter size at
farrowing. A healthy sow in good flesh needs
to gain 50 pounds less than a gilt during gestation — only 75 pounds. A nursed-down or
very thin sow, or a second-litter gilt, may need
to gain more than that additional 75 pounds.
Added feed or increased fat content in the feed
may be in order.

During the first two-thirds of gestation,
embryo growth is quite modest, and most sows
can get along quite well on a simple maintenance ration. This can often be met with just
4 pounds daily of a complete ration with a
crude-protein content of 14 to 15 percent.
They should be provided with about ½ pound
of crude protein daily. In very cold or raw
weather, the females might need an extra
pound of the complete ration or just corn to
help maintain condition.

In the last third of the gestation period, the
fetal litter makes its greatest surge of growth.
The demands on the sow's bodily reserves
increase, and these demands will continue
throughout the lactation period. During
this stage of the gestation, daily feed levels
may have to be increased by as much as 1 to
2 pounds. It is also the time to bring out the
highest-quality feedstuffs.

Prefarrowing Cleanup

Two to four weeks ahead of farrowing, you
should begin a cleanup process to ready the
females. This begins with a thorough treatment inside and out for parasites. See chapter
10 for more details on worming.

WATERING BREEDING STOCK

Most hogs today are watered with some
sort of tank- or vat-type waterer that maintains drinking water before the hogs at all
times. Kerosene burners or electric heaters are normally used to keep the waterers open in winter; painting them black
will help the tanks draw a bit of solar
heat. Offering your hogs all the water
they can drink in 20 minutes twice a day
will also work if there is plenty of trough
space and no animal is forced back.

The second stage of the cleanup moves to
the farrowing quarters. During my FFA days, it
showed up on our grade card when a surprise
visit by the ag teacher revealed dirty houses,
crates, or stalls. Before a new set of sows went
in, the farrowing quarters were to be cleaned
and scraped down to bare wood or metal. Then
they were saturated with a strong disinfecting
spray solution such as diluted iodine.

In an earlier day they would have been
sprayed or scrubbed down with a strong lye
solution or even smoke-fumigated. Why anyone would ever want to eat off a hog-house
floor was never made clear to us, but the grade
book made sure that we could.

Current thinking is that a simple cleaning
and scraping are generally more than enough.
Placing a couple of forkfuls of spent bedding
from the farrowing house into the gestation
pens a month ahead of sow due dates and getting the sows into the farrowing quarters a
week ahead of their farrowing dates will help
each animal develop her own natural immunity to any harmful organisms. This is also

SIGNS OF LABOR

As the time nears for the farrowing, the sow will give a number of clues to the imminent birth of her litter, although not all sows will show all signs. She:

- Will become restless

- Will have milk letdown

- May display nest-building behavior

- May have a slight vaginal discharge

- May circle about her pen and sniff behind her often

When you see these signs, the sow is about to begin heavy labor. Once farrowing starts, the sow will lie down on her side, and the pigs should be born at roughly 20-minute intervals. The smallest pigs will be born first and last in the litter. An old rule of thumb holds that small litters will come late and be mostly boar pigs; large litters will arrive early and be mostly gilts.

immunity that she can pass on to her young through the milk for the first few weeks of life. It is also far easier than trying to sterilize all the porous surfaces and nooks and crannies in a farrowing unit. Modern steam cleaners and pressure washers make cleanup of farrowing quarters even easier.

Farrowing

Get the sows into the farrowing quarters a few days early so they can settle in and develop a comfortable routine. Trying to move a sow too close to her farrowing time can leave her unsettled and greatly increase the number of pigs lost due to overlay or other fretful behavior.

The thought of taking a sow or sows through the farrowing process has probably put off more would-be pork producers than any other aspect of pork production. Actually, it is one of the simplest, most basic and natural of occurrences and generally happens with no need for excessive human intervention.

If you have selected large, well-grown females, fed them well, maintained their health, and matched the right type of boar to them, there is really very little room for problems. In 30 years at this hog game, I can recall just a handful of rough farrowings, and the veterinarian was able to talk me through a couple of them over the telephone.

Trouble Indicators

The sow's behavior and actions serve as good indications of something that's going wrong. Three or four hours of nonproductive labor or obvious signs of great pain or stress on the sow's part lets you know that intervention is needed. In most instances, a sow can safely deliver a breach or dead pig on her own, but this will slow down the birth process. The same is true of a very large pig, and if the pigs behind it are caught too long in the birth canal they can, quite literally, drown.

Other signs of a problem delivery include a sow that gets up and down often, a foul-smelling or bloody discharge from the vagina, and a pig's limb or head (especially with an extended or blue-appearing tongue) that has only partially emerged.

When problems become evident, the quicker a veterinarian is called in or other intervention made, the better. More pigs may be saved, the sow will fare better, and your costs should be less in the long run.

In some cases I have reached inside the sow's birth canal myself to assist with a birth or just to better determine how things are progressing. When this is necessary, there are a number of basic steps to follow:

1. Clip all the fingernails short on the hand you're going to use.

2. Thoroughly clean your hands and arms to well past the elbows.

3. Lubricate your hand and arm with lubricant purchased from your vet (mild liquid soap will do in a pinch).

4. Shape fingers and thumb into a sort of bird's beak.

5. Enter through the vulva, and proceed slowly until any obstruction is encountered.

There is not a lot of room within the birth canal, so you have to proceed quite slowly. When you encounter a pig, you must determine its position to assist in its birth. Grasping a pig in the birth canal can be difficult; there are a number of stainless surgical snares that can be sterilized and used for pulling pigs. Again, in a pinch, you can even boil a bit of baling twine and use it to form a simple obstetrical snare. If possible, try to grab the head; if it's a breech birth, draw the pig out by the hind legs.

Prolonged labor can tire a sow and greatly slow her contractions. You need to get the obstructing pig out quickly but not in a way that will injure the pig or sow. Try to time your slow and steady pulling to coincide with the sow's contractions, recognized by a tightening and drawing motion. Pull slowly and steadily and with a gentle angling downward toward her hocks.

Oxytocin is a prescription drug that can stimulate uterine contractions and milk letdown. It may also be one of the most abused drugs used in modern livestock production. Some use it just to speed up the birth process; others, without checking the birth canal for the cause of the slowdown. With a very large or dead pig blocking the birth canal, the drug may do little more than wear the sow down further. Also, it is easy to give an overdose of this drug. Our vet recommends just one or two injections of no more than 1 to 2 cc each.

Often, getting one pig moving will get the birth process going apace again. The pigs should then begin arriving at regular intervals, and after that, the **placenta** must be expelled. If the sow fails to "clean" — that's a livestock producer's term for shedding all the placenta— a severe infection can develop.

Anytime the sow's birth canal is entered, she should receive a substantial injection of a long-lasting antibiotic such as LA-200. I give 10 cc on each side of the neck. (Check chapter 10 for instructions on how to give injections.)

There is a very real question as to how much involvement you should have with a normally farrowing sow. A healthy female with good body capacity should be able to lie down and farrow at least eight live pigs, if a gilt, and at least nine if a sow. You can be a bit more forgiving with a gilt or a purebred litter, but

Young pigs resting comfortably. They are content and have good body fill, and the haircoat is smooth.

FARROWING NOTES

The small-scale producer is apt to farrow three to five sows in mid- to late spring and those same sows again in early to mid-fall.

Such farrowings can often be done on pasture or with simple sources of supplemental heat for the pigs for the first few days of life. The sows will be lactating and the pigs making that most important early growth during times of fairly favorable weather conditions and temperature.

With weaning at 5 to 8 weeks of age, such pigs will have "go anywhere" status. They will have received sufficient natural hardening to move to a feeding floor, into a dry- or woodlot feeding area, onto pasture, set to glean after harvest, or even go into a finishing unit. It is an aspect of versatility very much lacking in early-weaned pigs coming out of controlled-environment facilities.

The small-scale producer in simpler facilities should not be swayed or dismayed by the current high levels of pig output being touted for sows in controlled-environment facilities. When pork is reduced to a mere commodity, numbers of head pushed through a building to offset constant overhead costs take precedence over all other aspects of production, including the quality and desirability of the end product.

These supposed 30-plus-pigs-per-year sows carry heavy concentrations of white breeding and farrow smaller pigs at birth. The pigs are the result of terminal crosses, meaning all will be sent to slaughter as soon as possible, they are early weaned and often started with complex liquid diets, little or no mention is made of death rates following weaning, and energy costs can rival feed costs in their production. On the small farm with simple hog facilities, a six-pig litter at weaning will still likely cover all of the costs to produce in much the same way that it did 50 years ago.

Eight- and nine-pig litter averages are doable in simple housing and with purebred sows producing purebred pigs if careful selection has been made for good litter size at both farrowing and weaning. There are other swine traits with higher degrees of inheritability, but we were able to make much progress in this area by simply using breeding boars from litters of at least nine at both farrowing and weaning and with multiple littermates that grew out to demonstrate what might be termed "keeper" quality.

most of the females that miss this mark should earn a one-way trip to town.

Thirty years of careful selection for body capacity ensures farrowing ease. Rarely do our sows have trouble. I now bed our piggy females down following evening chores and generally don't go back out until morning. In short, with a normally farrowing gilt or sow that has been selected for body capacity, your involvement will be minimal.

The Pig at Birth

Most pigs are born with their eyes open, good reflexes, and a strong will to nurse and live. Many hog farmers stay by the sow's side to dry off the newborn pigs and place them on a teat. I have some most pleasant memories of long evenings in the farrowing house with Dad, the air rich with the smells of straw and his pipe tobacco, discussing the merits of each pig as it was born.

Young pigs nursing and showing the competitive behavior that can result in injury if the needle teeth aren't clipped

PROVIDING BREATHING HELP TO NEWBORNS

Pigs that needed extra time to be born or that were born surrounded by a membrane initially may need help breathing. There is a simple procedure to help them gain that all-important first breath.

- Clear the pig's mouth and nose of any fluids or membrane with your fingers.

- To remove fluids from the pig's mouth and lungs, suspend it head down, and slowly swing it back and forth between your legs.

- Again clear the mouth and nose.

- Dry the pig with clean rags.

- Set it down to nurse that nutrient-rich first milk.

- Monitor the pig's condition closely for the first few hours following its birth.

A watchful eye can save the occasional pig. Some are born in a saclike membrane; unless they are freed from it, they will suffocate. Sometimes you may encounter the odd female that becomes tense at farrowing and may **savage** (turn on and kill) her newborn pigs. Studies indicate that savaging of piglets may be more likely to occur in first litters and that the victim is more likely to be the firstborn. It also has been shown that sows that demonstrate this behavior were more likely to have been bred at low weight and in a depleted condition. Nevertheless, such females should be culled quickly and their offspring never allowed to enter the breeding herd because they may carry on this nervous behavior. Newborn pigs can be withheld from their dams for up to four hours following farrowing for their protection or be taken somewhere for warming.

Caring for the Newborn Pig

In the first three weeks of its life, a pig is a very different creature from the animal it will become. It is much more temperature sensitive and almost totally dependent on an all-liquid diet. Pigs with a birth weight of 3 pounds or more also seem to have a far greater chance of survival than pigs with lighter birth weights, especially weights of less than 2 pounds.

Little pigs will nurse about 16 times a day for several minutes at a time. Their very nursing activities stimulate milk production and release, and a contented sow will make soft, guttural noises while the pigs nurse.

Young pigs often develop diarrhea, also called **scours**, and can soon dehydrate and all too quickly die. At about three weeks following farrowing, a sow's milk flow peaks. It is about the same time that she would be entering her estrus cycle were she not nursing. Many believe that this may even key a chemical change in the composition of her milk. Such subtle changes can trigger scouring or increase stress in pigs not being well and carefully tended.

A swine disease called **transmissible gastroenteritis (TGE)** is especially insidious. Pigs with this condition may both scour and vomit, and the death rate in very young pigs can approach 100 percent. The fecal odor with a TGE outbreak is quite distinct.

TOO HOT OR TOO COLD?

Temperature extremes can be deadly to young pigs, so watch for signs that they are either chilled or overheated. Young pigs that are piled up and have their haircoats on end are chilled and huddled together for warmth (pigs at right, A). They are obviously under stress. If they are spread out and panting, they are too hot (pig at left, B).

More on scouring and TGE appears in the next chapter, but I'll say here that the best course of action to take with TGE is to counter the symptoms by rehydrating affected animals, keeping them warm, and allowing the next set of sows due to farrow to build natural immunity. If they are 30 days or more away from farrowing, they can be exposed to contaminated bedding and fed the contents of the intestines of pigs stricken with the disease.

The real trick is to keep such health problems away from the farrowing units and the breeding herd in the first place.

Colostrum

The all-important first milk — **colostrum** — imparts natural immunity that the young pig must have to ensure its very survival. In fact, it is virtually impossible for a pig to survive if it does not receive any colostrum.

Colostrum is secreted in the first 12 to 24 hours following farrowing. A pig needs to nurse within 4 hours of birth. A pig that isn't nursing will weaken and grow cold, and it may move away from the sow to a distant point in the crate or stall. A pig that survives the first day without receiving colostrum will fail to thrive and may die many days later.

Goat colostrum is probably the best substitute for a sow's, but cow colostrum will also work. Sow's milk may be the richest given by any domestic animal. Replacement colostrum can be frozen in an ice-cube tray; pop the frozen cubes into a plastic bag, and store them in the deep freeze for future use. Do not thaw the frozen colostrum in a microwave oven, however, or heat it with overly hot water; both can destroy vital **antibodies** in the colostrum, making it useless to the baby pigs. Thaw it in a double boiler or at room temperature.

Initial Management Tasks

Within 12 to 24 hours of farrowing, several management tasks must be addressed. These are listed below. Not every producer uses every one of these practices, but you'll probably want to employ at least some of them.

Grouping several treatments at the same time when the pigs are quite young minimizes stress. For instance, many hog farmers clip the **wolf teeth**, treat the navel, notch the ears, and dock the tails during a single session.

Wolf teeth clipping. There are eight wolf or **needle teeth** in the pig: two on each side of the upper jaw and two on each side of the lower. They should be clipped off at the jaw line. For this task, many hog producers use small clippers or sidecutters.

If they're not removed, wolf teeth can cause such discomfort to the sow's udder that the sow may not let the pigs nurse. Wolf teeth also can be used by the little pigs to cut each other in struggles over space at the udder. Pigs with small, crusty-looking scabbing on the snout and face are victims of needle-tooth lacerations.

The wolf teeth

INJECTIONS FOR NEWBORNS

On their first day of life, I give pigs a ½ cc injection of a long-lasting antibiotic to counter any potential problems with navel ill or other early infections. The injection site is one of the long muscles in the side of the neck. (More information on how to give an injection appears in chapter 10.)

In the other side of the neck can go a 1 to 2 cc supplemental-iron injection. A sow's milk is richer than that of any other common farm animal and would be the perfect food were it not lacking in iron content. Without that boost of iron, the pigs may soon develop a problem with iron-deficiency anemia. I and many other hog producers use a 1 cc injection initially, then give a second when the pigs are 10 to 14 days of age. And another iron injection may be in order at or shortly before weaning.

There are alternatives to injectable iron. Some iron products can be given orally via a pump. And there is a group of products commonly called "rootin' iron" because the mineral is in dry form, mixed with a peat moss carrier that pigs actually root through. I prefer the injectables, then the oral products, because they ensure that each pig gets an exact amount of the needed iron. For the "needle shy," the oral pump bottle may be appropriate.

Be sure to buy well-made syringes with fast, gliding actions. Before and during use, take the time to inspect needles for burrs and point-edge roughness (tossing any that are damaged), and replace them often. Give no more than three to eight injections per needle, inspecting them and cleaning them with the utmost care after each use. Many producers replace them after just one use.

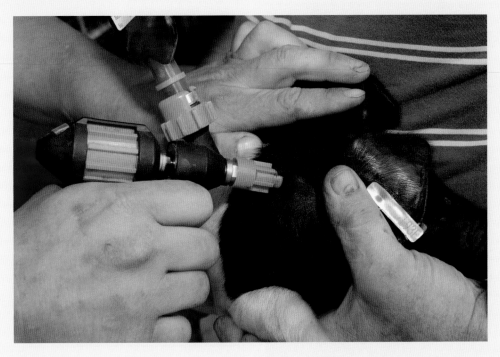

Here, a state-of-the-art injection system is being used in the farrowing house. Most small-scale producers will be well served with simple and inexpensive nylon syringes.

Navel treatment. Treat the navel with an astringent, such as an iodine solution, wound spray, or Blue Lotion. Along with cuts and scrapes at the knee joints, the navel is one of the primary routes infection can take into the very young pig.

Ear notching. Young pigs often have their ears notched, clipped, or punched as a means of individual identification. This is useful for age verification, as well as positively linking the pig to its sire and dam. Many producers also put a simple notch or punch a small hole in one ear of gilts from big litters that they wish to denote for future consideration as female herd replacements. If done in the first 12 hours following farrowing, ear notching is quite simple and causes the little pigs scant stress or pain.

The ear-notching method shown here is the one most commonly used and is required by most of the major swine registries for the issuing of pedigree documents. Some states may require this type of individual identification for certain disease-control measures and the issuance of transport documents.

Tail docking is now a somewhat controversial procedure, because animal activists charge that it causes pain. But in pigs, as in newborn dogs, the tail right after birth is simply soft gristle, and docking it appears to cause minimal, if any, discomfort. In fact, there is virtually no bleeding.

Feeder pigs are often sharply discounted at auctions if they are not tail docked. This is because pigs crowding in to a self-feeder or in fairly close confinement may engage in tail biting to move other pigs out of their way. Such activity can create sites for infection and, in extreme instances, has even led to cannibalism.

To avoid this, use sidecutter pliers or nippers to remove the last half or two-thirds of a pig's tail at the same time the pig is ear notched and has its needle teeth clipped. Clipping the tail much shorter than this may lead to problems with rectal prolapse, in which the rectum actually slips out of its proper place.

Many hog farmers will not dock gilts that are under consideration as female herd replacements. And we have had a problem selling boars that have lost a portion of their tail because it is considered cosmetically undesirable by some.

Right ear is notched to identify litter; left ear shows pig's individual number in the litter. The notches could be a little deeper.

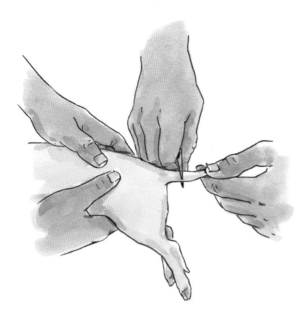

Don't clip the tail too short; it could lead to problems.

Castration

Boars not intended for stud service should be castrated because the hormones that emerge when they reach maturity could taint their meat. You'll want to castrate soon after birth, however, not as they're reaching maturity, because that's when the animals are easiest to handle and will heal the quickest.

Castration may be an even more controversial procedure than tail docking. Still, in this country the meat from even the youngest uncastrated boars is sharply discounted in price. In Europe, a lot of lightweight hogs are used for fresh and processed pork products, many of them intact males of 5 months of age or less. They are slaughtered before many of the more troublesome secondary male sexual characteristics can develop. These are triggered by male hormones and can range from coarse hair and the development of tusks to the boar taint that the packing industry so fears.

In the meat, it manifests itself as both a strong cooking odor and an unpleasant taste.

Over the years, we have eaten pork from a number of young boars. As long as they were not in service and were processed at a reasonable slaughter weight and age, we had no problems with the meat. Our local processor, however, will not render their fat for lard lest there be some trace of taint that might affect a whole kettle of lard.

The earlier a young male is castrated, the better, and the less stressful the procedure will be. Castrating a boar that has already been in service is pointless. Because I produce boars for sale as breeding stock, I have waited until as late as 10 weeks of age before castrating the last of a group of young males.

Castration procedure. Castration can be done through the **scrotum**. It works with pigs of all ages — although with animals of any size, it requires two people. Place the larger pig on its side on a table or straw bale; one person, serving as a holder, applies pressure with a knee or arm just ahead of the ham to hold the hindquarters very steady.

The necessary incisions tend to be rather small and exacting. Make a small cut over each testis and through the scrotum; draw the testis free and sever it along with a bit of its supporting cord; then treat the surgical wound with an antiseptic astringent or wound protector.

To make the incision, you can use a single-edged razor blade, a pocketknife with spay blade, a scalpel, or one of the new disposable, hooked-blade castration knives (about $1 each and adequate for 10 to 12 castrations if properly used and dipped in alcohol or cleaned after each use). An aerosol-spray antiseptic is very useful for treating the surgical wounds, because it penetrates them quite deeply.

HUMANE OR NOT HUMANE?

It is not my place to take sides in animal rights issues in this volume, but events in recent years have had some effect on certain animal-care practices and how they are perceived by those outside producer circles.

Some of these concerns have grown out of misunderstandings and the failure of producers to better maintain lines of communication between consumers and themselves. Most producers, myself included, frequently have given ourselves over to the care and safe maintenance of our animals, and we have the scars to show for it.

Many of the practices that have given rise to concern actually grew out of frontline concerns for the animals' well-being and producer and animal safety and wanting to increase animal survival rates. I have seen littermate boars raised together simply turn on each other at puberty and fight to the death. One of the stockperson's roles is to temper nature's law of the fang and the claw and to provide proper animal safeguards whenever and wherever possible.

All the early pig procedures listed in this chapter are certainly made better by the use of proper methods, are skills that the producer has mastered or those that are practiced by the professionals the producer engages, are done with clean and well-maintained equipment, are done in a timely fashion, and are carried out when animals are as young as possible.

Little pigs younger than 5 days of age have rather simple nervous systems, have high thresholds of pain, bleed less, and tolerate most basic health-care procedures quite well. Leave needle teeth unclipped, however, and they will quickly create sores on the sow's udder. Baby pigs with wolf teeth intact cut each other when nursing. And even the smallest of cuts is a potential infection site.

In the overall production scheme, a good health-care kit is quite inexpensive to assemble and maintain. Items within it should be replaced as soon as they show signs of wear or damage. A number of items are now made to be used only once as a health safeguard and to ensure sharp points or edges for fast, clean usage.

One of the more controversial practices now in livestock care is castration. It is a subject that is painful for many people to discuss, but there are some very sound reasons for the practice.

Currently, market hogs are being fed to heavier weights and may be on the farm to 170 days of age or more. And intact males may develop secondary sexual characteristics, including aggressiveness, at a fairly early age. If left intact, they are more apt to fight with each other, begin breeding activity to the stress of other young penmates, become more difficult to handle, and be harder on housing and fencing. They are also prone to ranting behavior. This is self-debilitating sexual aggression that includes pushing fences, hissing, frothing at the mouth, mounting behavior, failure to eat, and just generally behaving like an 18-year-old boy. As mentioned, after a certain age, intact males can impart to the meat that they yield certain unpleasant tastes and aromas — the boar taint.

Some have advocated administering anesthesia before castration, but it is expensive, requires special skills and equipment, can be a somewhat protracted process, and can increase death loss — hogs are one species that do not deal well with anesthesia. In some countries, the castration practice is gradually being done away with, but there are also countries where hogs are harvested at younger ages and at lighter weights. Much of the pork in those countries is used in sausage and other highly flavored, processed items. On farms where castration is not used, young females and males are sexually segregated at an early age and fed out in separate pen groupings.

Over the years, we have eaten a number of young, intact males, although we always had them processed at 160 days of age or younger. Even then, however, our local processor would discard their fat trimmings rather than possibly introduce the boar taint to a whole lard rendering. The meat was quite good, but much of it we did have made into sausage or pork burger.

Chemical castration is on the horizon and represents a group of products many may not wish to see used with meat-producing animals.

The castration process and most other practices that draw even the smallest amount of blood pose a conundrum for the times. The pros, to date, outweigh the cons. Consumers will certainly have more to say about them in the long term, but they may have to back up their concerns with a change in pork-consuming habits and a willingness to pay higher prices. Castration is a practice common to all the major livestock species, and when asked to speak on poultry topics, I now encounter a growing interest in the caponizing process for chickens.

Human civilization has marched shoulder to shoulder with the practice of keeping livestock. You cannot have an organic agriculture without manure and spent bedding to return to the soil to replenish it. Only when people stopped the practice of simply following after the wild herds could they begin the business of building cities and civilizations.

The role the keeping of livestock has had on humans is shown in everything from cultural food favorites to idioms of human speech. The tending of the flocks and herds is a calling tinged with dust, sweat, and, yes, blood. It always has been and always will be.

The need for some of these early procedures may be reduced with changing consumer patterns, but the consequences of some of them going undone can create pain and a risk of death or injury that is as great as that of the procedure or even greater for the animal and possibly its penmates.

There is another method of castration, and with it one person can actually castrate a pig of up to 60 pounds or so. It is sometimes called the "show barrow cut," because it leaves no visible scarring. The method calls for a pig holder, a device that slips onto a gate top or stall partition.

Slip the pig's hind legs through members on the holder, which suspends the animal head down and with its belly facing out. The testes move downward between the animal's hind legs. Place a knee between the pig's front legs to hold it steady, and grasp one testis at a time between thumb and forefinger to bring it taut against the skin (this causes no pain).

Make a short incision with a hooked castration blade just above the testis. The pressure you are applying beneath it with the thumb and forefinger of your other hand should cause the testis to pop through the incision. Draw it out, along with a goodly portion of the cord-type system that supports it. Then sever the cord to free the testis. You can treat the wounds at this point with a good antiseptic spray.

Because castration requires strict attention to detail and can be challenging, it would be wise to enlist the help of your veterinarian or an experienced hog producer to teach you the castration procedure.

Castration aftercare. Freshly castrated pigs should be returned to a dry, freshly bedded stall or pen. There should be no mud for them to lie in, and water should be offered in pans or troughs in which the pigs cannot lie. This will keep water and foreign substances from entering the surgical wounds.

The pigs should be watched closely for several days for signs of infection, such as excessive scrotal swelling or fever. I have never had a

problem following castration; the wounds heal and close quickly.

Sometimes, however, a hernia results after castration. One of the animal's testicles will drop lower than the other, appear larger, or otherwise be asymmetrical. In this case the intestines must be returned to their proper place and the site sutured, so you may well need the help of your veterinarian.

Postfarrowing Sow Care

On the day of farrowing, many hog farmers totally withhold feed from the sow, the idea being to circumvent gastric upset or constipation, which might disrupt milk flow.

For a few days before and after farrowing, some also add bran to farrowing rations as a further preventive measure or **top-dress** the ration with oat bran or beet pulp, which comes in a meal-like form. Some of our Amish friends offer warm water with Epsom salts in it for the first drink following farrowing.

It is important to make note of when a sow cleans, rises, and begins normal bowel function following farrowing. Getting her up is critical. A few sows just flat out fail to rise for what seems to be no really good reason — perhaps a fear of a wet, slippery floor — and this may be one of the few times when the use of an electric prod is justified. She is well past farrowing, has cleaned, is not mothering, and may be showing heat stress. You must get her up following farrowing. If you don't, she won't eat, and she'll eventually be lost to a combination of stress, shock, pneumonia, and pure cussedness. I want my sows up as quickly as possible following farrowing — within one to two hours.

Virtually from the moment of farrowing, you must monitor the sow for signs of the

MMA complex: mastitis, metritis, and agalactia. A uterine infection may cause a fever to set in, the udder to redden and harden, and the milk flow to decline or even stop. Be on the watch for vaginal discharge, reddening and stiffening of the udder, and excessive squealing or gauntness in the pigs, which indicates the sow is nursing poorly and the pigs aren't getting enough to eat. The condition may actually become so painful that the sow will lie on her belly to keep the pigs from nursing, causing her further discomfort.

Normal treatment procedures for MMA are antibiotic injections, uterine infusions, and oxytocin, sometimes called "pop," to stimulate milk letdown. Just remember not to abuse the latter.

Feeding the Nursing Sow

In most instances, we feed our breeding animals just once a day, in the late afternoon. In summer, the cooling temperatures at that time seem to stimulate appetite, and in winter the sows go into the long, cold night with full stomachs.

However, the demands of nursing a litter require increased feed levels for the sow. From 12 to 24 hours following farrowing, many producers offer the sow about 2 pounds of her normal ration; then they gradually build the amount fed over the course of a week or so until the sow is on full feed (about 3 percent of her liveweight). One old rule of thumb holds that within a week or so of farrowing the sow should be receiving 3 to 4 pounds of feed a day for her own needs and an additional 1 pound for each pig she is nursing.

Some nursing sows may need as much as 16 to 20 pounds of feed per day in the latter stages of the **lactation** period. Certainly, feed to appetite. That old rule of thumb goes on to advise 4 to 6 pounds for the sow during late lactation and 1 pound for each nursing pig. Generally speaking, the more you can get into the sow at this time, the better.

Many hog producers still divide up the sow feed and give it to the breeding animals twice a day, morning and evening. To get enough feed into lactating sows with large litters in warm weather, three-times-a-day feeding may be in order, especially when it's very hot — and then even three feedings may not be enough to get a sow to consume all she really needs. The heat suppresses metabolic functions as well as appetite. You can also add fat to the ration at the rate of 3 to 5 percent of the total amount fed, but you must do so gradually to prevent gastric upset. Some producers even top off the sow's daily ration with 1 to 2 pounds of milk and nutrient-rich pig starter (discussed on page 220) if she is nursing an especially large litter.

The nursing sow should also receive free-choice drinking water. Not only are demands being made on her by her nursing young, but she must also maintain body condition adequate to allow for a rapid breedback following weaning. Females that are neither nursing nor pregnant are costing their owner money. A hollow sow is a mighty expensive creature to own.

MILK PRODUCTION

A variety of factors, such as general body condition, can affect milk production. But the most important factor is good nutrition. Make sure your sows are on a proper diet to help ensure that they produce enough milk.

Care for Nursing Pigs

The next major decision in nursing management is how and when to provide pig starter — a ready-made, store-bought product — or even if it is really necessary to provide such complex feed if the sow is nursing well. Pig starter is perhaps the most complexly formulated of all swine feeds: It is high in both fat and protein content, is richly flavored, and has a substantial level of milk product. It is also the most costly, and rations for very young pigs, or **prestarters**, are often still more complex and costly. There is even a category of liquid pig starters and ration boosters designed to closely approximate sow's milk.

Trying to get the various starter feeds to the nursing pigs can also be a bit of a challenge. I have seen sows in crates literally turn backflips or stand on their heads to reach creep feed if they can. Those little metal and plastic-crate pig feeders don't last very long when a sow gets to them.

In the 1960s, pigs were weaned at 8 weeks in the United States. Before that, it was left up to the sow. I recall one noted Midwestern Yorkshire swine breeder who pioneered the weaning of pigs at 35 days of age. It was his contention that with 35-day weaning, if a sow was worth her salt at all as a pig raiser, there was really no need for pig starter to be offered to the nursing pigs. He called it little more than expensive pig bedding.

There are now multiple-stage pig starters intended to be fed to nursing pigs at various stages of their life, beginning just a few days after farrowing. Some are even meant to be fed at the rate of no more than a couple of pounds per head before changing to the next feed in the program. The greatest justification for such feeds is in early-weaning situations — weaning the pigs at 10 to 21 days of age.

With later weaning — at 6 to 8 weeks — a starter/grower feed can be useful in maintaining growth in the pigs, because the sow's milk production begins to decline three to five weeks into the lactation period. These simpler feeds still pack a bit of milk and flavoring but typically have crude-protein levels in the 16 to 18 percent range. Some of these early starters and prestarters crowd 30 percent crude

SOURCES OF STRESS

Throughout this book, I emphasize the importance of minimizing stress on pigs and hogs, because stress lowers animals' resistance to disease. Sometimes, of course, it is not possible to completely avoid stress — such as the stress of weaning. Still, always do whatever you can to minimize it. And there are other types of stress that can be prevented entirely, such as overcrowding. Here's a list of the various stressors that can affect swine:

- Crowding
- Inadequate feeding or watering space
- Temperature changes
- Introducing new pigs
- Feed changes
- Weaning

Keep in mind that hauling and medical procedures needn't be stressful if done correctly and with care.

protein and have such exotic ingredients as spray-dried blood product.

The walk-in creep feeder made of wood and sheet metal has been in use for decades for later-weaned pigs, especially those in group situations on drylot or pasture. In theory such feeders admit only the pigs, blocking out the sows that can consume many dollars' worth of the pricey feed in just a few minutes. In the real world, though, some old sows can play these feeders like a hustler plays a pinball machine. For best results, gate off such feeders when in use to protect them from the sows.

On average, a young pig will consume 40 to 50 pounds of starter/grower, along with mother's milk, when growing from birth to 40 or 45 pounds. By 2 weeks of age, the pigs will also be robbing a bit from their mama's feedstuffs.

Weaning

As noted, pig weaning now sometimes occurs when the pigs are as young as 10 to 14 days of age. Quite a departure from the traditional 56-day weaning age! Over the years, I've also seen some pigs left on the sows until they were as much as 10 weeks — 70 days — old.

I found that 56-day weaning took a greater toll on our purebred sows than I really liked to see and also made it rather difficult to attain five litters farrowed within a 24-month period. Thus I tried 42-day weaning, then dropped down to weaning the pigs at 35 days of age. Weaning too early — less than 21 days post-farrowing — has been found to cause problems for some sows with rebreeding. It seems to run almost totally counter to any natural cycles.

Our 35-day-old pigs are good size, have been venturing away from Mama and their houses for some time, are swiping regularly

from the sow's feed, and are hardened off and acclimated to outside air temperatures. They can handle the less complex, more moderately priced starter/growers and have been raised by their dams long enough to give us a good indication of the sow's milking and mothering abilities.

Weaning Options

Seas of ink have been expended on discussions of housing and maintaining the newly weaned pig. Very-early-weaned pigs have to be moved into buildings capable of maintaining rigidly controlled environments. There, they are kept in temperatures often exceeding 80°F (27°C), fed elaborate rations (sometimes even flavored with chocolate or strawberry), and held in single-litter pens, coops, or tublike structures.

Two hot buzzwords these days are SEW and MEW. **SEW** stands for **segregated early weaning** and is a measure taken to break swine-disease cycles. The pigs are weaned when they are only about 10 pounds, which is quite young. They are moved to a nursery facility far removed from any other hogs — often a facility all alone on another farm several miles away from the rest of the hog herd. The idea is that distancing the young pigs from all other hogs keeps them from exposure to disease via these hogs or the environment that surrounds them.

MEW denotes **medicated early weaning**. Also a practice intended to break disease cycles, it was developed in the United Kingdom and modified for use in this country. Basically, sows are immunized against several diseases before farrowing to enhance colostral immunity, and weaned pigs are moved to an isolated nursery unit, where they are separated from pigs of other ages. The separation by age and the isolation appear to be more effective in controlling disease than medicating them, according to Drs. Ross P. Cowart and Stan W. Casteel, authors of *An Outline of Swine Diseases* (Wiley-Blackwell, 2002).

Both SEW and MEW are quite costly practices to pursue. They are really outgrowths of the high-volume approach to swine production — the factory farming of pork. In such operations the swine-raising facilities are in virtually constant use, large numbers of animals are held in a very small area, stress loads on the animals are far greater, and individual animals don't receive adequate attention from the producer. The animals are relentlessly pushed to keep production on a factory-like schedule and to keep the facilities full. All of this sets the stage for near-constant struggles with the "swine disease of the month."

On the small farm, however, where hogs are a part of a diversified enterprise mix and weaning ages run later, nursery accommodations can be far simpler. In fact, the least stressful way to wean pigs is to pull the sows and leave the pigs where they've always been. They are most at ease and comfortable in those surroundings. Note that your fencing must be quite pig tight to contain the pigs during the weaning period. They will walk the pen perimeters steadily looking for an opening that will lead them back to Mama. The sows should also be moved beyond scent, sight, and earshot of the pigs to further reduce stress on both groups. They will miss their pigs at first but quickly return to herd-animal status and to rebreeding.

Feeding the Weaning Pig

Weaning can be one of the most stressful periods in a young pig's life. For at least one week before and after, there should be no changes made in its diet. When changes do occur in a pig's diet, they should be gradual, over a period of three days or so. In the first feeding, offer a mixture of about one-third new ration and two-thirds current ration (which is the pig starter or a second-stage starter/grower feed). On the second day, mix the new and old rations half and half. On the third day, blend two-thirds new ration with one-third old. Pigs are, by the way, free fed: offer as much as they want to eat.

Early growing rations generally have a crude-protein content of 15 to 18 percent, and if bought by the 50-pound bag come in the form of small pellets or crumble. They are highly palatable and easily digestible. It is possible to formulate early grower and even starter rations on the farm, but some of their components have a quite short shelf life and are expensive to buy in small lots.

The cost and complexity of these rations is generally justified by the rapid and very efficient way in which young hogs grow. They can be the most positive investment you make in swine feedstuffs.

Growout

With the stress and trials of weaning behind it, a pig begins its real work in life — making a hog of itself.

When pigs are about 10 weeks of age, hog producers begin blending litters. Feeding/finishing hogs are manageable in groups of up to 60 head. On most small farms, the finishing groups typically contain the pigs from between 2 and 10 litters — 15 to 75 pigs per group.

MANAGING PIGS AT WEANING

At 5 to 8 weeks of age, newly weaned pigs are not exactly hothouse orchids, but giving them a few breaks will do much to ensure that they get off to the best start possible. Some steps to help them through this potentially stressful time include:

1. Try to hold the newly weaned pigs in litter-size or penmate groups for at least the first few days following weaning.

2. Tie or solidly prop open lids on self-feeding and watering equipment, and closely watch over the pigs for several days following weaning. Newly weaned pigs may have trouble operating this equipment. It's normal for them to lose a bit of bloom or have a growth standstill for a short time, but watch closely for any that are becoming gaunt or losing tone or vitality.

   We once sold a set of just-weaned pigs to someone who hoped to save a few dollars by buying them young. He took them home to a large wooded lot with a creek running through it and a single, small walk-in feeder. The pigs found the water, but they had no experience with this type of feeder, nor were they familiar with the type of feed he was using (a complete meal rather than a pelleted grower). Several pigs actually died before the rest mastered the new feeder and ration. A call to our vet confirmed that this is an all-too-common occurrence.

3. Because pigs will drink when they won't eat, an inexpensive packet of vitamin/electrolyte powder mixed into their drinking water will help recently weaned animals maintain health and body condition. A box of flavored gelatin powder added to the drinking water will increase consumption levels.

4. For the first few days following weaning, it may be best to offer feedstuffs and drinking water in shallow pans or troughs. Feed presented this way can be changed often to keep it fresh and appealing.

SELF-FEEDER SPACE FOR FINISHERS

The rule of thumb is to provide one self-feeder lid or hole for every three to five finishing hogs in a pen or group. Most such feeders hold 40 to 80 bushels of feed (based on the 56-pounds-per-bushel weight of shelled corn) and have 10 to 12 feeder lids. Producers generally try to match even-sized hogs, pen, and feeding and watering equipment to create a system that functions simply and requires only a modest amount of care and supervision.

Feeding to Finish

Hog finishing can at first seem to be little more than keeping the feeder and waterer full and watching the hogs grow, but there is more to it than that. Often much more.

Feeding. A finishing practice growing in acceptance is **split-sex feeding**. This practice takes into account the different finishing traits of the sexes. Gilts grow a bit more slowly than barrows and tend to be leaner. They use protein to build muscle. Barrows grow faster but tend to lay on more finish (fat). As they grow, protein levels can be throttled back a bit in their rations.

Today's hogs are generally a lot better in type and lean content than those of even a decade ago and can do a first-rate job of utilizing high-performance rations. From about 60 pounds to market weight, our Durocs and their crosses stay on a 15 percent grower — a pretty hot feed. It helps our animals realize their optimum potential while not breaking the bank when we go to the elevator.

Never neglect drinking-water quality and freshness. It is every bit as important as quality feedstuffs.

Observing. Farmers don't get up with the chickens any more, but watching finishing hogs come off their beds in the morning can be most helpful. This is a good time to spot respiratory ills, for instance, which often reveal themselves as coughing spells when the hogs first exert themselves. It is also a good time to spot injury and soundness problems. As hogs come up, their underlines and extremities are very visible. Those last pigs off their beds need an extra-close inspection to determine the cause of their sluggishness.

Providing adequate space. Be sure to provide the growing hogs with plenty of sleeping space. They need 8 square feet per head at minimum, but 10 may be more in order. The animals need room to spread out in warm weather and plenty of dry bedding in cold weather. In the latter case, chilled hogs will pile up, and prolapses or even deaths by crushing and suffocation can result.

With small pigs, it may be necessary to block off a portion of the sleeping area until they grow larger. This is to prevent them from picking up the habit of dunging inside their sleeping quarters.

Handling topenders and tailenders. Seldom do all the hogs in a group mature and reach handy market weight all at the same time; to be honest, I don't think I've ever seen that happen. You may need to **top out** a pen, which means removing hogs as they reach good market weight, two or three times over a period of weeks. This prevents your top-end hogs from becoming too heavy and receiving a price dock for carrying excess finish.

In nearly every pen there will be a tailender or two. These are pigs that, due to injury, health problems, pure cussedness, or whatever, grow slowly or even stall out. A lot of them seem to hit the wall at 170 to 180 pounds. You can drop them back to another group of younger finishers, but they seldom perform efficiently, and you can put yourself through a lot of stress and strain getting them blended into another set of hogs. These are good candidates for the home freezer or a trip to the local sale barn. They will sell for less, but it is often far better and simpler to just move them on and get your pen empty.

This model of self-feeder has no trough lids to facilitate the pigs in locating the feed and using the feeder. It should not be used outside of a building.

THINK BEFORE YOU SPEAK ABOUT YOUR HOGS

This may seem like a small thing, but we producers need to do a better job of how we refer to our growing hogs and finishing pen. A lot of time and dollars have been spent to make hogs leaner and more efficient, and a lot of this is belied every time a farmer is heard in a public place talking about "fat hogs," "fattening hogs," or the "fattening pen." These terms still persist because many don't like newer terms such as "harvestable," and the term "mature" doesn't quite fit. An old term I like is "ready to go to town."

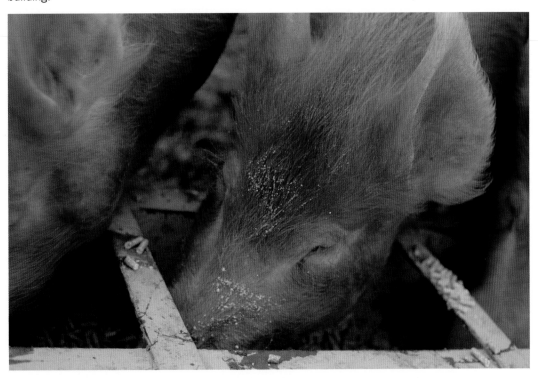

10

HOG HEALTH CARE

The hog is one of the most vigorous, most prolific, and hardiest of all domestic animals — that is, if it is allowed to remain true to its nature. Care of a small swine enterprise takes a modest block of time each day; it is largely a matter of keeping the hogs comfortable and content. When hogs are provided with good, basic care, they will generally have minimal health-care needs.

Problems do arise, however, as with any animal species. The longer I live and work with livestock, the more I have come to appreciate the old axiom, "You can think and breed your way out." This is true of a great many potential health problems. Hogs have not endured these last 6,000 years or so because they were frail and easily sickened creatures. In most cases, they are compromised only when efforts are made to bend them to fit a set of production parameters that are totally unnatural and enormously stressful.

A New Vulnerability

In the confinement methods characteristic of megaproducers, hogs have had to contend with a stifling environment and a place of steel and concrete (hard edges and unyielding surfaces), are seen as inputs rather than animals, and are compromised by those always willing to settle for the quick fix. In these environments,

health ills come at the animals with the speed and volume of machine-gun fire. When *A Guide to Raising Pigs* was first published, the big swine health-care concern was pseudorabies, a disease that had been around in one form or another for decades.

Now they and their producers must contend with all sorts of syndromes and complexes, such as Porcine Reproductive and Respiratory Syndrome (PRRS). They are health ills that manifest in all sorts of performance- and reproductive-disrupting states and with varying degrees of actual death loss. The confinement sector is, quite simply, trying to force hogs to live and perform in an environment almost as alien to them as if it were the far side of the moon. In fact, these practices are the way you would have to raise hogs on the moon!

The hog is a woodland creature. If left in a more natural state, fed well, and kept hydrated, it can cope with varying temperatures, dampness, and pretty much all that the natural world can throw it. That includes bears and wolves.

However, start paring off all traces of fat cover, breed for unusually heavy muscling, wean at a very early age, pen them on hard surfaces, cause them to live near or adjacent to pools of their own wastes, deny access to sunshine and fresh air, feed the lowest-cost rations (that change often), and leave them in an

Shoats on pasture

environment that's stark even by nineteenth-century prison standards — and can there be any wonder why health problems emerge?

We are almost to the point where we have two different porcine species: One is the true hog and the other is the confinement animal. Some lines have now been selectively bred for decades for life in confinement, and they are largely disasters outside of that environment.

With the growing demand for more naturally and humanely reared pork, there has been a recent scramble for seedstock for newly established outdoor herds. The problem is that there is now a rather limited supply of these breeds, and confinement-bred hogs will not work outside. I was on an antibiotic-free farm in a range setting awhile back and the two herd boars, both obviously confinement bred, had pus discharging from their eyes in the chill of a late-March Missouri morning.

Pick up certain environmental and news magazines now or watch some of those generally negative film clips on the evening news, and you will get a pretty good view of this new creature: big eared, whitish, somewhat dirty in appearance, and often worn looking. It's a tough life; most sows don't survive past a fourth parity (litter) in confinement, and they are small, with little salvage value. These hogs are more energy dependent than a luxury SUV because they have limited fat cover and live on concrete surfaces and under a roof where they need supplemental heat, lighting, cooling, and ventilation.

Of course all mammals are at some risk for disease regardless of their environment. And some of the diseases befalling a species are more prevalent than others. When a major livestock disease is arrested, it often clears a space for another, once lesser health problem to break through and perhaps even grow in consequence. What once held diseases in check naturally were the long-time practices of holding livestock in modest numbers and in more widely dispersed herds or flocks, and careful selection and breeding for factors most conducive to health and well-being. There is no measure for body capacity, soundness, and will to live beyond the stockperson's eye and experience.

We got off the tracks pretty badly with the quest for not so much a good meat hog but a 300-pound walking slab of pork to be made into riblets and restructured, flaked pork products. The first key to good hog health was set down long ago with the admonition from veteran producers that hogs needed to walk on dirt, have sun fall on their backs, be given the food to do the job, and receive encouragement to be just what God meant them to be — hogs!

Breed Your Way Back to Vigor

According to poet Gertrude Stein, a "rose is a rose is a rose," but no longer are all hogs the same. The small producer hoping to establish or add to a more naturalized, outdoor-based herd will be badly served by today's confinement-bred and -reared hogs, very badly served.

Not long ago at a farm auction, I was asked by a group of young Amish farmers how to deal with an ileitis problem that had emerged in their modest-sized, antibiotic-free operations. When you choose to produce without antibiotics, the diseases and parasites don't just go away; your stockpersonship had better get good.

Their herds were young and built largely from animals drawn from the confinement sector. I had no pat answer for them, no "one shot" solution. I actually may have troubled them more than provided them with the immediate help they wanted, but I did receive an appreciative nod from an older stockperson who had been standing nearby, a sort of Amish high five, if you will.

My answer to them was to dig in their heels and get busy breeding their way out of trouble. The best health plan ever created was set down by Mother Nature long, long ago; it is simply the natural order — screening and selecting for viability, durability, and survivability. We were 35 years with a Duroc sow herd and didn't get it anywhere close to where we wanted to be until we were a good two decades into it.

When we sought to add a second breeding line, we went through six different gilts before finding the one that fit our program and mated with our herd boars and worked in our simpler facilities and open air lots. A hog with the best performance data in the world isn't worth a dime to any producer if it dies, goes lame, or is lacking in **libido**. Nor is it good business to allow the überperforming hog to live and breed on if it must have all sorts of drug and environmental crutches.

It is a painful thing to pay hundreds of dollars for a gilt and then slap her on the rump and send her to town. Still more painful and costly, however, is cursing her and all of her offspring on your farm years later because you didn't make that first sacrifice. Experience has taught me that before anything else you must select for basic vigor and what we have come to call stoutness. You want them big, strong, full of vinegar and prunes, and ready for whatever life may throw at them. And that begins in the first hours of life in the farrowing pen.

Planning Ahead for Health

You must limit your selections to litters from females that lie down slowly, have their pigs quickly, and mother them well. Those pigs,

in turn, have to rise up quickly, make it to her side, and fight for that udder segment — showing that all-important will to live. There are now all sorts of testing methods and performance indexes for the so-called economic traits (growth rate, feed efficiency, loineye size, etc.), but nothing has yet or ever will replace the stock raiser's experienced eye when it comes to evaluating an animal's vigor and simple "do-ability" merits.

While you can't exactly breed for good health, you can breed for better health. Vigor and the will to live might seem to be esoteric topics, but they do manifest themselves in some very real and economically important ways. To be blunt: sick hogs don't grow well, and dead ones produce no young.

I have never advocated selecting for extremes in type, but if I could impart only one bit of management planning for improved health and vigor, it would be to select replacement gilts only from the largest and fastest growing third of the pig crop and boars only from the most rugged and growthy, top 20 percent of their pig crop. Skim that cream time after time, and you will build a good herd for your particular farm environment. I have seen swine feeds tweaked more than some racing car fuel formulas, but the fuel works best only when it is supplying a motor that is built tough.

Deep, wide chests and big, deep body cavities say that the animal is able to take in and process ample amounts of breath, feed, and water. A large frame is what it all hangs on, and heavy bones transport the hog from feed to water to shelter. You can't produce the performance data to document it, and to inexperienced ears it can create an unsettling image, but among veteran producers, "He's a pretty tough ol' hog" is still high praise.

BUYER BEWARE

Salesmen, farm-supply stores, catalogs, the Internet, and other sources put before us all sorts of health supplies and treatments. Some are good, some are ho-hum, some are designed to expire soon after you buy them, and some are flat-out illegal to own and use. For many years, one of the great rural folk legends was about a little shop behind the stockyards in Kansas City or St. Paul or Chicago where an old gentleman had a magic-bullet cure for this livestock disease or that one.

It was a good story, and there are some veterinarians who can and do prescribe off-label uses for some products. Some years ago, a certain turkey health-care product was found by producers to work on a quite major swine malady. It was never cleared by the government for that use, yet large displays of the product could be found in feed stores all across hog country.

Do be most careful and discerning when it comes to health-care practices and products. Your best money as a producer will be spent on quality feedstuffs and safeguarding animal comfort. In the worst of weather, I knew I had done the best by my hogs if I left them well fed and watered and with clean, dry bedding before the coming night.

A Health-Care Philosophy

As I mentioned in chapter 3, I believe hog producers must avoid the temptation to become drug and medication happy. Hogs and pork are considered in many corners to be over-medicated, and the industry as a whole is perceived to be heavily drug dependent. There is at least partial truth to both charges, and this has hurt sales and consumption of pork and pork products.

In nature, when the population of a given species becomes too great for a certain locale, natural forces intercede to trim the population to more sustainable numbers. Generally, these take the form of some sort of contagion or infectious agent. The same is true when domestic animals are packed too tightly into an artificial environment.

On farms with hundreds or thousands of head of hogs, there is just no way to break disease and parasite life cycles. There are always pigs that are abnormally small, injured animals, stunted and stressed pigs, vulnerable sows giving birth, and sick pens that never stand empty. These are never-ceasing reservoirs of sickness and infection, often existing right next to sows constantly giving birth.

I have an acquaintance who manages a two-thousand-plus sow operation in Illinois, in which many of his efforts and those of his 10-person staff go toward health care. Newborn pigs are actually given drug and vitamin infusions through stomach tubes. There are far too many swine farms now weighed down with such multiple-injection programs — not to increase performance but to simply stay online with at least some semblance of production.

Fewer May Be Better

I am neither an organic producer nor a health purist, but I do find that my workday and bottom line are both made better by breeding our hogs to be tougher and more durable rather than by buying animal drugs by the case lot. Producers with modest numbers can be truly discerning users of what is on the market. They don't have to jump on every new product down the pike just to stay ahead of the disease fires burning down in the hog lots.

The livestock industry appears to be on the verge of losing a number of current animal health products, has seen a number of others become available through prescription only, and may be shut out entirely from the use of any new-generation products that may also be usable by humans. This is because there are concerns about resistance developing to products if they are overused, and I think these concerns have some validity.

Due to the sheer numbers of animals — and health problems — on some farms, many producers now have to medicate constantly to keep those problems in check and reasonably manageable. Alas, their solution is being imposed upon a great many other producers whether they need it or not. How? Read on.

Drugs and steroids have kept a lot of hogs in the gene pool producing when they should have been washed out of the herd long before reaching the breeding pen. When these bad ones breed, so do their problems. Consumer confidence in pork will grow only as a commonsense approach to hog health care is applied all across the industry.

The counterpoint to subtherapeutic antibiotic use is management on a human scale to maximize the strengths and natural hardiness of the hog. In contrast to the large operator, small hog farmers have time to give to their animals, and that's what will ensure they are never driven out of business by the corporate operators. They can give attention to even the

VETERINARY FEED DIRECTIVE

Recent measures to reduce the use of antibiotics in livestock feed require a closer and more structured relationship between producer and veterinarian. Most vets now want more knowledge of the operation and more farm visits than the infrequent emergency call. They ask how the animals are managed, their breeding and age, the purpose for which they are intended, and how and where the treated rations are to be prepared.

Small producers have not generally been the group abusing antibiotics; they try to pursue a simpler and more natural course of production, and their animals are not held in an overly stressful environment. A balance must be struck and consumer wishes respected, but let's not take a needed tool completely out of the hands of those who know best how to use it, or create circumstances that could push some of the smallest producers out of business by adding too greatly to their operating costs.

smallest details, can even be about the business of fine-tuning the enterprise, and have a commitment to quality, harmony, and product wholesomeness that will always keep their production in demand.

Reading about the diseases reviewed later in this chapter will drill this point home. It will become apparent that many swine illnesses can be prevented through management — the type of management that the small producer can provide.

FIVE GENERAL TIPS FOR PREVENTING DISEASE

Here are five ways that you can manage an operation for improved hog health:

1. Stagger farrowings. If you don't have too many little pigs around at the same time, your workload will be reduced, enabling you to give attention to important animal health-care details.

2. Practice lot and pasture rotation. This helps to control parasites and mud, which will keep the animals more comfortable.

3. Have empathy for hogs, and provide extra time and care to those that require it. On most small farms every sow is known on sight, and her history comes quickly to mind.

4. Allow facilities to stand idle and open to the cleansing rays of the sun.

5. Employ selective breeding for vigorous and durable hogs. Select for deep, wide chests; large body cavities; high birth and weaning weights; large, loose frames; good mothering; well-formed genitalia; and obvious health and vigor.

Hogs will respond in kind to the treatment that they receive.

Just the Basics

I wish I could set down for you in these pages a full and complete guide to swine health, but to do that this book would have to run to several thousand pages and then be updated weekly, if not daily. In this chapter is an outline for a small-farm health-care plan and brief descriptions of many of the more common swine ailments.

The basics of diagnostic skills are outlined here, but be forewarned: In their early stages, many health problems mimic each other. And certain regions of the country may have a history of health concerns not found in others. For example, we are in an area with a history of kidney worm problems in swine, a parasite not widely seen in other, even nearby areas. The producer must therefore endeavor to be fully informed about health matters occurring both locally and nationally. And when in doubt, always seek the input of a professional. It can be costly to consult with a veterinarian, but sometimes just a simple phone call can save time, dollars, and possible loss.

One of the best ways to save money on health care is to learn how to administer many medicines and treatments yourself. It can also often save precious hours, which can be the difference between life and death for hogs. If the veterinarian is aware of your skill level, he or she can often make recommendations that can even result in further cash savings. Our vet once walked me through a troublesome birth over the telephone.

Of course, we all have our limits. My grandfather loved his animals dearly. He would hunt and dress game and help butcher them; however, he could never bring himself to give injections or draw blood. Some feel that they lack the size and strength to treat large animals,

while others are very comfortable with the work. We had a neighbor that supplemented his income by doing basic health practices like docking lambs and castrating calves and pigs.

Do that with which you feel comfortable, work hard to build your level of expertise, read extensively in the field and stay current on health issues, build a good relationship with your veterinarian and approved health-product suppliers, and never fall for the old saw that if a 10 cc dose is good a 50 cc one has to be better.

Staying Current

To achieve good swine herd health, it is critical that you stay current with what is happening both locally and nationally. This is one thing you can do yourself with the right resources. We have weathered many local disease outbreaks ranging from transmissible gastroenteritis (TGE) to pseudorabies in part by being plugged into the local scene. Through it you learn which local communities are experiencing disease outbreaks, which delivery trucks not to allow on the farm, which used-equipment sales not to attend, and sometimes when to just stay home, period.

Six different swine-oriented magazines arrive in our mailbox monthly, and we also receive a newsletter from our area's Extension livestock specialist.

Veterinary Care

We have cultivated a relationship with our veterinarian that enables us to pick up the phone and get answers that quiet our fears about rumored outbreaks and give us firsthand information on the latest in animal-health products.

To have a similar relationship with your vet, you have to call him or her for more than just tending midnight emergencies and salvaging animal wrecks. Our vet is on the farm several times each year for routine procedures such as blood testing, he knows our herd and our management practices right down to the brands of feed we use, and he knows me and my abilities. I also patronize him for a number of health supplies. For instance, he was the one who steered me to the speed syringe (see page 236).

Granted, his prices for products may be higher than those in some of the mail-order catalogs. Still, his products are always fresh and without close expiration dates, he can show me how to use them, they are available from him in small amounts, and they are just minutes away when needed.

SUPPLY KIT

Every producer with hogs needs a medi-kit of supplies. I keep mine in a small plastic carrying tray grooved in the bottom to fit solidly on the top board of a gate or stall divider. This tray and its contents also fit neatly into a secondhand refrigerator we keep in an outbuilding to store animal health products at the correct temperatures.

The supplies my kit contains are shown in the list below. You'll learn about the use of these products in the rest of this chapter.

- An assortment of disposable syringes in sizes ranging from 3 to 10 cc

- An assortment of needles in their sanitary plastic guards (used needles can be kept in a small screw-top jar filled with alcohol)

- A surgical-blade handle and package of hooked and straight surgical blades (for tasks such as lancing abscesses)

- An aerosol wound protector or spray-type product (the latter does not have the hissing sound that many hogs find so upsetting)

- A marking crayon for identifying treated hogs (greens and yellows are the most visible on hogs)

- A rectal thermometer

- A broad-spectrum injectable antibiotic such as LA-200 (a form of Terramycin)

- An antibiotic product for use in the hogs' drinking water

- Epinephrine for shock

- Both a snare and tongs for restraining hogs

- Prescription products, as directed by the veterinarian

- A package of a vitamin/electrolyte powder (which can be added to the drinking water to counter stress)

- An over-the-counter scours (diarrhea) remedy

- A rehydration product

- A good standard animal-health text such as *The Merck Veterinary Manual*

Injections

We have raised literally hundreds of boars that received absolutely no other injections than their iron shot at birth. Still, injections are necessary to some extent, and they are one of those health-care skills that even the smallest producer should acquire, to both trim operational costs and provide better care for his or her charges.

Besides the iron shot for newborns, you may have to provide antibiotics if your hogs experience a disease outbreak. Waterborne antibiotics are more appropriate for support therapy, but when you need a fast-acting antibiotic, you'll want to inject it.

The labels on the products you use for injection will advise you about the route of administration — be sure to follow these instructions. There are three common injection methods for swine.

Subcutaneous (SubQ), which means just beneath the animal's skin. SubQ shots usually are given in the shoulder/neck region.

Intramuscular (IM), or into a muscle. This is by far the most common method of injection in hogs. With the other two methods — subcutaneous and intraperitoneal — absorption of the medicant is slower but more sustained. With intramuscular injection, there is rapid absorption, because muscles are well supplied with blood.

Intramuscular injections are given into large muscles that lie just beneath the skin. In times past, the prime target for IM shots in hogs was the ham muscle. However, there is an increased pain factor with injections in that region; some injections in the ham muscle have caused nerve damage and even lameness. The better target is one of the long muscles on either side of the neck. These work

for newborns right on up to the largest sows and boars. They're along the animal's **topline**, making them easily reachable; they're accessible even on animals held in a catch gate; and they're a good site choice for nearly all injectable health products.

Intraperitoneal, or within the peritoneal cavity, which is where loose connective tissue covers the surface of tendons and ligaments. Intraperitoneal injections go into hollows adjacent to body cavities. Perhaps most typical of this method is the injection of epinephrine to counter shock. On a shoat, it can be given in one of the underleg areas that correspond to the human armpit or groin area.

Common site for administering a subcutaneous or intramuscular injection

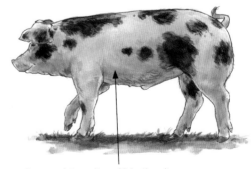

Common intraperitoneal injection sites

HOW TO GIVE AN INJECTION

Giving an injection to an animal that's both larger than yourself and very agile can be a daunting task. My grandfather, who paid our local veterinarian to teach me how to give injections and do castration work rather than do them himself, used to argue that the harder it was to inject a given hog, the less likely it was that that animal really needed an injection. If they can run across 40 acres, they may not be as sick as you thought.

Our vet is able to get up to 20 cc of an injectable product into a completely unrestrained sow with the following method:

- Use new needles to be assured that they are absolutely sharp and free of burrs. The few cents it costs to change needles is nearly always money well spent.
- Use one of the new disposable syringes or the speed models, made of nylon or plastic, that work with a squeeze of the hand. We have a 10 cc version of the latter that fits in the palm of the hand.
- Load the exact dosage needed, because the first injection is the only easy one.
- Use the shortest and heaviest-gauge needle you can — the thickest, in other words — or it may not penetrate the skin. To inject sows or boars, I prefer a 1-inch-long, 14-gauge needle. For small pigs, I prefer a ½-inch-long needle that is 18 gauge.
- Use a long-lasting product to reduce the number of injections needed for a full course of treatment. Some antibiotics now remain at strong levels in an animal's body for as long as 48 to 72 hours.
- Slowly and quietly approach the animal from the rear, holding the syringe parallel to the animal's topline.
- When the syringe is over the injection site, use a fluid motion to tilt the syringe up, drive the needle home, and depress the syringe.
- A pinch or hand slap on the site as the needle is withdrawn will prevent any serum leak from the puncture wound.

Personally, I have found that the two keys to making injections easier are the simple nylon syringes — which stroke smoothly and quickly — and clean, sharp needles of the shortest length and heaviest gauge possible.

As I have indicated, the giving of injections is a very useful skill, but if no amount of written instruction can give you the confidence you need to try it, enlist the help of your veterinarian. Chances are the vet will gladly give you a lesson, because if you can give injections in the future it will clear his or her calendar for more important diagnostic and health work.

Parasite Control

Hogs contract a variety of parasites, which can adversely affect weight gain and, in some cases, make pigs sick. Some parasites, such as roundworms, are internal; others are external parasites, such as mites, which affect the skin. For more information on specific types of parasites, see Swine Ailments and Illnesses, beginning on page 239.

Early methods of countering internal-parasite buildup, or worms, depended almost totally on regular rotation of various swine lots and pastures to break up worm life cycles; each lot was used just once a year. Some rather noxious and dubious concoctions were also fed or given in the drinking water to try to knock out worms.

Today we have more effective options for worm control, but no one product seems up to the task of removing all the parasites that can affect hogs. That's why it's important to rotate the parasite-control products you use. Further, there is strong evidence that repeated use of the same worming product may cause some parasite species to develop a resistance to it.

Administering Parasite-Control Products

Some worming products can be given by injection, others can be mixed into feed (treated feed is the only ration offered to the hogs over a period of one to three days), and still others are mixed into drinking water. Products for external parasites, such as Ivomec, may be injected, but they also come in other forms, such as pour-ons. The surest way to worm is by injection, because you are assured that each animal gets the exact dosage for its weight and age.

It is more difficult to administer wormers in water, especially during cold weather, when water usage is reduced and there is the risk that the water will freeze before all the wormer is consumed. And if you administer wormers in the feed, there is always the concern that the smaller pigs in the group or pen that can't compete with larger pigs won't get enough.

For control of the more common external parasites, lice and mange mites, you can choose the injectable Ivomec or any of a number of spray, pour-on, and dust-type products. Your selection will be dictated by both the environment and the season of the year.

TYPES OF WORMERS

In no way do I wish to appear to favor one product over another here. Many worming products are useful, even some of the older ones such as piperazine. Still, at Willow Valley we rotate the products Ivomec, Atgard, and Tramisol.

With Ivomec, which contains the **anthelmintic** ivermectin, we control not only a wide variety of internal parasites but external parasites such as mange mites and lice as well. Ivomec comes in an injectable form as well as a pour-on formulation.

Atgard kills many internal parasites at all stages of development. It is a granulated product that must be mixed with mealtime rations and should not be mixed with pellets. You can buy it premixed.

Tramisol is a powder that is mixed with water, then added to drinking water. It kills lungworms, which rely on red earthworms as an intermediate host and can be a special problem in hogs living outside or in wooded lots.

PARASITE-CONTROL SCHEDULE

It's a good idea to check with knowledgeable hog producers in your area or your local swine veterinarian before selecting parasite-control products. These people can help you determine which products will be most effective for the parasites you need to control. However, you do want to rotate worming products to help ensure that parasites don't become resistant to one specific kind. Try not to use the same worming product more than twice in a row.

Here is an example of a parasite-control plan based on the way we do things at Willow Valley:

New pigs brought onto the farm. Seven to 10 days after arrival, treat with injectable Ivomec or Tramisol. Then administer a second treatment of Ivomec or Atgard when the animals reach 150 to 185 pounds.

In the breeding herd. Treat all breeding animals for external parasites. If it's warm weather, a liquid pour-on may be best; if it's cold, a powder. Injectable ivermectin is a good option but may be costly.

Sows. Two to four weeks before farrowing, treat sows for internals and externals. If you did not use injectable Ivomec last time, it would be a good option now. Otherwise, this time you might want to treat with Tramisol and use something else for external parasites, such as one of the pour-on or powder products.

Young pigs. We worm at 6 to 8 weeks, about one week after weaning, with Ivomec injectable.

The sprays and pour-ons are most effective for external parasites, but they have two drawbacks: Some cannot be used too close to farrowing, lest traces of the product be ingested by the nursing pigs; and in weather below 40°F (4°C), problems with chilling and stress may arise from the wetting of the hogs.

Lice problems increase in cold weather, when the hogs lie more often in the bedding and sleep closer together. A good pour-on can be applied along each hog's topline on a pretty fall day, and dust-type products can be used on the hogs and bedding in cold weather. The dust-type products may be among the least effective, however, due to reduced contact times; still, they may be helpful in the coldest of weather.

Be aware that there is a withdrawal time for some parasite-control products. You cannot use these products for a certain number of days before the hogs go to market; for some products, it's as long as 75 days. Be sure to read the label on each product to determine its withdrawal time.

Alternative Parasite-Control Products

An organic parasite-control method is the feeding of and dusting with diatomaceous earth. This is a component of a product that we administer to our hogs called Immuno-Boost. When we offer this product in the drinking water along with another worming measure, it does seem to increase the hogs' response to the wormer.

I have never used exclusively diatomaceous earth for worming and have doubts about just how effective it can be, but I do think it gives some results. Anyone doing organic hog farming should also be diligent about rotating pastures to help control parasites.

Swine Ailments and Illnesses

One chapter in a book cannot begin to cover all the illnesses and ailments that affect swine, but here you will get some of the basics.

There are a number of problems in swine that you'll be able to manage yourself; for others, you'll need to enlist the help of your veterinarian. Most important, however, is that you familiarize yourself with the signs of various important illnesses in your hogs. If you have any doubt about what's wrong, or are concerned that you may not be able to handle the problem yourself, consult your veterinarian. In hogs, prompt treatment of diseases can be lifesaving and also prevent a slowdown in weight gain.

Abscesses

Abscesses are pus-filled swellings caused by bacterial infection. They are most noticeable when right under the skin and can range from thumb to fist size. Some hog producers lance the abscess with a clean blade and apply hand pressure to facilitate drainage, but this procedure can introduce the abscess-causing organism to the rest of the animals in a pen. If you lance, do it away from other animals. Afterward, apply a good common drying astringent, such as Blue Lotion or an iodine product.

Treatment from "within" is the more common approach. Abscesses normally respond to commonly used injectable antibiotics.

Atrophic Rhinitis

Atrophic rhinitis is a mouthful of a disease generally just called rhinitis. It is a respiratory condition that is thought to be due to infection by more than one "bug" and is transmitted to young pigs probably via nose-to-nose contact with their dam.

Atrophic rhinitis results in inflammation in the mucosal lining of the nose. Early symptoms include sneezing, sniffling, possibly a bloody discharge from the nose, and a watery discharge from the eyes. In more advanced stages the bones in the snout (**turbinates**) may atrophy, and the snout may even swell and twist. The disease can slow growth dramatically, and as turbinate damage progresses, the hog also becomes far more vulnerable to pneumonia.

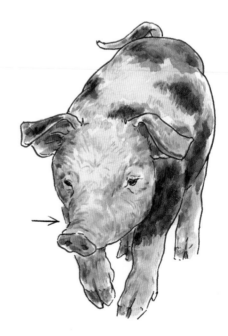

In the advanced stages of atrophic rhinitis, the bones in the snout (turbinates) may atrophy, and the snout may even swell and twist.

In the very early stages of disease, rhinitis may be undetected in white breeds because the eye and nose discharge may not be as noticeable on white faces as on dark.

A preventive injection—a vaccine—exists that can be given to sows before farrowing and to young pigs in herds with this disease. It is usually combined with antibiotic treatment and changes in management procedures. However, once the vaccination program for this disease is initiated on a farm, it must be continued.

At one time, it was thought that as many as 85 percent of the hogs in the United States had at least a touch of atrophic rhinitis. Animals sure to be free of rhinitis and some other diseases are those taken from their dams by cesarean section and raised in isolation; they are called specific pathogen free (SPF) pigs.

Although some of the symptoms are similar, this is not the same disease as necrotic rhinitis or bullnose in young hogs (see page 247).

VACCINATIONS

With good selection, care, and feeding, I have found that vaccinations against most respiratory diseases are seldom necessary. The problems we've had were the kind that, generally, we could treat as needed. However, some herds develop persistent diseases for which vaccination may be beneficial. If you have a herd that develops an infectious disease, discuss a vaccination program with your veterinarian.

Brucellosis

Brucellosis is a bacterial disease that is declining in significance in the United States following many years of doing blood testing on breeding herds and certifying them free of the disease before breeding stock from them is allowed to be sold.

Its most common symptoms—which can cause great economic loss — are abortions, stillbirths, and the birth of very weak pigs that fail to thrive. Brucellosis also causes an infection of the testicles in males and can lead to infertility. When testing reveals carriers of this disease, the animals are ordered destroyed. Undulant fever is the human form of brucellosis, and it once took its toll on farmers and animal-health workers.

Clostridial Enteritis

Clostridial enteritis is a bacterial gastrointestinal disease usually caused by an organism called *Clostridium perfringens* Type C. Veterinarians, especially in the Midwest, have reported an increase in the incidence of this disease in recent years.

Clostridial enteritis is usually seen in pigs' first few days of life, but it can affect weaned pigs, too. It can cause diarrhea that may or may not be bloody, and sometimes the disease is so virulent it causes sudden death in piglets before you realize anything is wrong. It is thought that the pigs contract the disease from the sow and from sow feces; the organism is also found in soil.

Clostridial enteritis can be very difficult to distinguish from other gastrointestinal diseases, especially those caused by another organism known as *E. coli*.

The treatment for clostridial enteritis is generally a course of antibiotics; injectable and feed-grade antibiotics also are often given, and veterinarians usually advise management procedures, especially beefed-up sanitation efforts aimed at reducing the spread of disease to baby pigs. This includes washing the sow before farrowing. Antitoxins and vaccines also have been used. It can be difficult to get clostridial enteritis under control once a herd is infected.

A less severe but chronic "Type A" form of this disease has also been identified. Infested pigs might develop mild diarrhea but not always, and the Type A disease can cause reductions in weight gain and feed efficiency. At least one veterinarian has reported that the Type A form has, in some herds, increased preweaning deaths among pigs that develop diarrhea; in pigs that survive, it can reduce weaning weights. Treatment may be necessary if the disease is causing problems, but in some herds this milder form of clostridial enteritis clears up on its own.

Dehydration

Dehydration is usually caused by illnesses that result in scouring, or diarrhea. Oral glucose and **electrolytes** are antidotes for dehydration, and there are a number of good ones, such as RE-SORB, available just for pigs. In a pinch you can use Pedialyte, a product for children available at your local drugstore, or even Gatorade, which can be administered orally with a syringe (without the needle attached). Keeping dehydrated animals warm and comfortable is also a good idea. The real trick is keeping health problems away from the farrowing units and the breeding herd.

E. coli

E. coli is a bacterium that can cause several illnesses. Diarrhea after weaning, for instance,

---

10 SIGNS OF ILLNESS

Throughout this book you have learned about the signs of illness in swine. Here is a roundup of 10 of them; all indicate you could have an infectious disease problem. Suspect trouble if animals are:

1. Slow to get off their beds
2. Eating less or not eating at all
3. Looking drawn
4. Suffering from diarrhea
5. Acting depressed
6. Vomiting
7. Experiencing abortions or stillborns
8. Lame or walking stiffly
9. Acting uncoordinated
10. Among nursing pigs, seeming to be "poor doers"

---

known as postweaning scours, has been associated with E. coli (although this problem has also been linked to early weaning). One of the more serious conditions linked to E. coli is hemorrhagic bowel syndrome, which is a highly fatal, massive hemorrhage into the intestines. Signs of this are lethargy and discomfort.

If your pigs break out with any serious infectious disease, consult your veterinarian immediately, to determine the cause and find the right treatment.

Erysipelas

Erysipelas is an old-line disease that still pops up from time to time to take a toll on hogs. It is a bacterial disease with symptoms that may include loss of appetite and fever; a stiff gait and arched back; the classic red diamond-shaped patterns appearing on the skin; and sudden death with no apparent symptoms.

Both injectables and a waterborne product are used to treat the disease.

Gastric Ulcers

One cause of gastric ulcers is thought to be giving hogs feed that is too finely ground, although factors such as stress might also play a role. Hogs with gastric ulcers may act dull and go off their feed. If their ulcers bleed, they may become anemic.

To prevent ulcers, be sure to give your hogs feed that is more than 3.5 mm screen size and keep them comfortable to reduce stress.

Glasser's Disease

The clinical name for Glasser's disease is *Haemophilus polyserositis*. That's because it's caused by the organism *Haemophilus parasuis* (*H. suis* for short), and it involves the joints as well as a generalized illness. The *H. suis* organism is one that hogs can harbor in their respiratory tracts without experiencing any problems; however, during times of stress, such as weaning or temperature changes, it can make pigs sick.

Glasser's disease develops suddenly and kills pigs quickly. It tends to strike pigs younger than 10 weeks of age. There is also a chronic form of the disease that causes all kinds of symptoms, including a fever, trouble breathing, swelling in the joints, and lameness.

This is a disease that tends to be more severe in hogs that are **specific pathogen free (SPF)**; those are pigs delivered by cesarean section and free of many diseases, such as mange. However, when SPF pigs are introduced into conventional herds and have no immunity to *H. suis*, they can become very ill with Glasser's disease at any age.

Treatment needs to be administered as quickly as possible and consists primarily of antibiotics. There is also a vaccine that a veterinarian can recommend if *H. suis* becomes a problem in your herd.

Greasy Pig Disease

Greasy pig disease is nothing more than a staph, or *Staphylococcus*, infection. If pigs have external parasites, such as mange, or traumatic injuries to the skin that are not treated, the infection can set in. Improper nutrition is also thought to be linked with the disease.

Greasy pig disease tends to affect pigs either just before or right after weaning. The skin flakes off, becomes crusty, and may crack. There may be a stinky dark brown secretion. Pigs may act dull and go off their feed. This disease can kill pigs if not treated.

The remedies are antibiotics, washing the pigs with medicated shampoos, and, sometimes, dipping them in solutions containing products such as chlorine bleach. To hasten healing, pigs should be treated for external parasites if they have them, and measures should be taken to see that their environment is comfortable and their feed adequate in nutrients.

Hog Cholera Eradicated

Hog cholera used to be dreaded in swine. It was a highly infectious viral disease that came on suddenly and had a high death rate. But thanks to an eradication program initiated in this country in 1962, hog cholera has been wiped out. It is still a problem, though, in several other countries.

Hyperthermia and Hypothermia

Hyperthermia means the body is overheated. I once lost a good gilt to hyperthermia in the middle of a snowstorm. A border collie puppy chased her around until she got overheated.

A hog that gets hyperthermia goes into shock. Do whatever you can to calm it. Get it away from anything that might upset it. Do not hose the hog down, however. If you cool down a hot hog, you'll have a dead hog on your hands. At most, brush a bit of cool water across the bridge of the nose. Administer an intraperitoneal injection of epinephrine, and keep your fingers crossed.

Hypothermia, or low body temperature, means the animal is too cold. It isn't a common problem in hogs. Little pigs, however, are susceptible to chilling, and keeping them warm is important. Be sure to read up on ways to do this in chapter 9.

Infectious Arthritis

Infectious arthritis is an infection that strikes the joints. Lameness is the most obvious sign

TAKING THE TEMPERATURE

The chart below, adapted from the text *Diseases of Swine*, 7th Edition (Iowa State University Press, 1992), shows the normal temperature for pigs of different ages.

In many other types of animals, it's common to check the heart and respiratory rate as well as the temperature. But in pigs and hogs, I've found that the temperature, coupled with the way the animals act, is a pretty good indicator of illness.

What does the temperature mean? As you'll see below, the hog's normal body temperature is about 101 to 102°F (38.3 to 38.9°C). When a fever develops, it's because the body's systems are rising to its own defense to fight off an infection — which it may or may not be able to do. But a below-normal temperature can indicate trouble, too. It can mean the body systems are shutting down, which occurs with kidney failure in swine.

When I take a hog's temperature, I use a standard rectal thermometer with a ring in the end. With a short piece of stainless wire I attach an alligator clip, which I then attach to the animal's haircoat. (This leaves my hands free for other chores while I wait.) I insert the thermometer about 3 inches into the rectum and leave it in for at least two minutes.

Pig Age	Rectal Temperature
Newborn	100.2°F (37.9°C)
1 hour old	98.3°F (36.8°C)
24 hours old	101.5°F (38.6°C)
Unweaned pig	102.6°F (39.2°C)
Growing pig (60–100 lbs [27.2–45.4 kg])	102.3°F (39.1°C)
Finishing pig (100–200 lbs [45.4–90.7 kg])	101.8°F (38.8°C)
Gestating sow	101.7°F (38.7°C)
Boar	101.1°F (38.4°C)

of infectious arthritis in the pig; it may affect only one limb, but it can also affect more. In newborn pigs, it may be due to exposure to pathogens in the uterus; after birth it can result from navel ill, or pigs might contract infections during tail docking, castration, and other management procedures. These procedures create points of entry for infection if the sites are not treated and carefully monitored. A number of organisms may be the culprit, including *E. coli*, *Streptococcus*, and *Staphylococcus*.

In pigs about 4 weeks of age or slightly older, infectious arthritis may be due to an organism called *Mycoplasma hyosynoviae*. This is an organism found in the nose, which then moves to the joints.

Some forms of infectious arthritis respond better to antibiotics than others, and some medications only work if given early in the course of disease, so you'll need a veterinarian's help to determine the exact cause of and best treatment for the problem. Avoiding the conditions that result in infectious arthritis is a better approach. This could be concrete flooring, kneeling at feeding equipment, or any other situation that causes damage to knees and hocks.

Iron Deficiency

I've advised more than once in this book that supplemental iron should be given to baby pigs. Without it, they are at risk for developing baby piglet anemia, or iron deficiency, because baby pigs require more iron than they receive from their sows. Baby pigs can become very anemic within a few days of birth. It's one of the very few things that's lacking in near-perfect sow's milk. In nature, pigs are born on the ground with an immediate source of soil-born iron sources.

Pigs with anemia won't grow normally, and they act dull. Their skin is pale, and they may scour or even have trouble breathing. They may also have a longer-than-normal haircoat.

If the initial dose of supplemental iron is injected, rather than orally administered, you can be sure the pigs receive what they need.

Leptospirosis

Commonly known as "lepto," leptospirosis is most often caused in swine by the infectious organism *Leptospirosis pomona*. The organism can be transmitted through the urine of infected animals, which may be dogs, cows, or rats as well as swine. It enters the body through the mucous membranes or breaks in the skin. Lepto can cause a fever, loss of appetite, and diarrhea. But its most common and costly effects in swine are abortion, stillbirths, and the birth of weak pigs.

Antibiotic therapy under the direction of a veterinarian is the recommended course of action after the cause of any abortions has been determined. There's also a vaccination for lepto, but the immunity it provides may be short-lived.

Lungworm

Hogs contract the lungworm parasite by ingesting infected earthworms. Larvae migrate to the heart as well as the lungs.

The result of lungworm infestation is similar to that of some stomach worms: Hogs may have trouble breathing and may develop an asthmatic cough characterized by a "thumping" noise. They may be "poor doers," animals that are growing slowly.

Good sanitation and an effective parasite-control program are the ways to control lungworm.

LICE AND MANGE

Lice and mange are common external parasites of swine.

Lice. Suspect lice if your pigs rub themselves to relieve itching, and especially if you see small white eggs on their hairs — which are lice nits. You can see lice on a pig, too, around its head, ears, and flank area. The remedy is use of a product that kills lice.

The pig louse (*Haematopinus suis*) is more likely to be seen in herds that are not well managed. It is transmitted via pig-to-pig contact, and the lice actually suck blood from the pig's skin, which can result in anemia. The primary problem that lice cause, however, is skin irritation and itching. Infested pigs may grow slowly and don't efficiently utilize their feed.

Mange. Like lice, mange is not uncommon among swine. It is an infection caused by a mite called *Sarcoptes scabiei*. This external parasite spreads via pig-to-pig contact, and it can be transmitted from an infected dam to her litter.

Mites favor the head and ear area. Hogs with mites will rub to relieve the itching and may develop scabs. Some young pigs are allergic to mites, so their response to this infection will be more severe; their itching will be worse, and skin irritation can involve their hind ends and abdominal areas. Young pigs with mite infestations grow more slowly.

The remedy for mange mites, as for lice, is treatment with a product that kills the parasites. Be sure to read Parasite Control, beginning on page 237.

louse

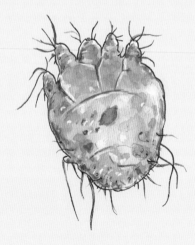

mange mite

Mastitis, Metritis, and Agalactia (MMA)

MMA is a syndrome characterized by udder infection (mastitis), uterine infection (metritis), and loss of milk complex (agalactia). It can strike sows shortly before or after farrowing. Sows may lose appetite, have a discharge from the vulva, run a fever, appear depressed, have a hardening of the udder, or even dry up completely. In short, a sow isn't able to provide enough milk for her litter. She may also become constipated, and her mammary glands may be enlarged and feel warm.

This syndrome can be caused by several problems, which include infections, stress, and dietary management problems. It is more common in sows that have had more litters.

Sows that develop MMA may require antibiotics and oxytocin to stimulate milk flow. Something to encourage appetite might also be in order: You might try pig starter or gelatin powder to stimulate their curiosity. There is also a product called Spur designed for this syndrome. Vitamin B injections may do the trick, and we've had good results by administering LA-200, an injectable antibiotic.

I've heard about giving sows everything from dog food to beer to stimulate appetite, but my concern there would be that such a change in diet might cause more problems than it has resolved with resultant gastric upset and hyper diarrhea.

The best approach to MMA is prevention. Add bulk to the sow's diet, keep her comfortable and clean, and provide her with opportunities to exercise.

Mycoplasmal Pneumonia

Mycoplasmal pneumonia is another mouthful of a disease, caused by an organism called *Mycoplasma hyopneumoniae*. It is thought by some to be the most common cause of chronic pneumonia in swine. It can be spread in the air and via direct contact.

This form of pneumonia is more likely to be seen in pigs at least 3 months of age. They develop a chronic, dry cough, which might not be evident until they exercise. They may or may not have a fever or lose their appetite, depending on the seriousness of the infection. This disease, like many others that affect swine, can stunt growth.

Mycoplasmal pneumonia may clear up on its own if pigs haven't been stricken too severely, but the problem is that when pigs have this disease, it often makes them susceptible to

TRACKING TREATMENTS

When you are administering treatments, such as antibiotic injections, to several animals, it can be difficult to keep track of which ones you've treated and which ones you haven't. To resolve this problem, I mark each animal as I administer the treatment. If some animals require more than one dose, I use two different marker colors: for instance, a red marker as I administer the first treatment, a green marker for the second, and so on.

For little pigs I've found that marking on the top of the head ensures that I see my color coding. I mark the big ones anywhere on the top line and with stock crayon that contrasts with their haircoat.

other infections, such as *Streptococcus suis*. In such cases, antibiotic treatment is necessary; there is also a vaccine available. This disease, too, is associated with overcrowding and poor sanitation, so avoid such stressful conditions.

Navel Ill

Navel ill is a general term used to describe an infection that starts in the umbilical area and spreads. The navel, in fact, is one of the primary routes of infection into the very young pig. You can prevent infections at the navel by applying an astringent, such as an iodine solution, wound spray, or Blue Lotion, from 12 to 24 hours after birth.

In pigs that develop navel ill, antibiotic treatment is probably necessary. Consult your veterinarian.

Necrotic Rhinitis (Bullnose)

Necrotic rhinitis is an uncommon disease, and I mention it here only to make it clear that it is different from atrophic rhinitis. Necrotic rhinitis is a bacterial infection that generally occurs on farms with poor management and sanitation. It might develop if management tasks such as nose ringing or wolf teeth cutting are performed under less than hygienic conditions or after young pigs cut their wolf teeth.

It's obvious where the common name of the disease — bullnose — came from: infected pigs develop serious swelling and abscesses of the snout. Improved sanitation practices and antibiotic treatment are used to resolve the condition.

Porcine Reproductive and Respiratory Syndrome (PRRS)

PRRS is a viral disease that results in several problems affecting — as the name implies — the reproductive and respiratory systems. It is thought to be spread via pig-to-pig contact, but airborne transmission is also possible. PRRS is considered one of the newer diseases, first reported in the late 1980s. Initial outbreaks occurred in Indiana, Iowa, and Minnesota.

Swine affected by PRRS may develop a fever, lose their appetite, and appear dull. The respiratory part of the syndrome causes the classic "thumping" type of breathing and sets the stage for bacterial infections to take hold. This breathing is characterized by heaving, rasping, a rattling sound, and a shaking body. Sometimes PRRS results in a lack of oxygen that turns the ears blue, a condition called blue ear disease. Abortions are another significant problem linked with the disease. PRRS tends to clear up after about two months but can sometimes be fatal.

To study the disease, the government has been working with hog producers whose herds are affected by PRRS by taking blood and tissue samples. Currently, there is no specific treatment for PRRS, and the best way to avoid it is to prevent your hogs from exposure to herds that have been sick with the syndrome. PRRS may also be more likely to occur in herds that have been exposed to fumonisin mycotoxin, a toxin formed from a fungus that can occur in feed.

Pseudorabies (Mad Itch)

Pseudorabies is an infection is caused by a herpes virus, and it is a disease that often gets attention in swine circles. To help hold it in check, most states now mandate a testing program for producers who sell breeding stock. It can affect hogs of all ages, but its symptoms quite often first appear as abortions and weak pigs at birth. Weaned and growing pigs with the disease develop pneumonia, a fever, poor coordination, and convulsions. Pseudorabies is usually fatal in pigs that aren't yet weaned; it's somewhat less deadly in weaned and growing pigs. Adult hogs with pseudorabies may have pneumonia or fever — or show no signs of infection at all.

Other farm animals as well as dogs and cats can contract the disease. All tend to experience intense itching, hence the name "mad itch." Pseudorabies is virtually always fatal in these other species.

There is no treatment for this disease, although antibiotics are sometimes used to treat the resulting pneumonia.

Extensive testing and destruction of infected animals is under way to control this costly disease. There is an injectable preventive that may reduce clinical signs of the disease but does not prevent the spread of infection in the herd. There's another problem with this product: Once it is used, your animals will test falsely positive for pseudorabies, and that can be murder if you are in the breeding business, because breeders want to impress upon buyers that the herd is disease-free. I thus recommend not using the preventive medicine unless the problem with pseudorabies is really serious in the herd.

Rotaviral Enteritis

Rotaviral enteritis is a virus that is thought to affect many herds. It may result in diarrhea among newly weaned pigs, and it's one of those diseases to suspect when nursing pigs are "poor doers." Rotavirus is often found along with other swine infections, such as E. coli.

There are vaccines available, but their results have been mixed, because they may not tackle all the strains of rotavirus that are affecting a herd. This disease requires good management of weaned pigs by providing them with a stress-free, comfortable environment and good diet. Treatment of bacterial infections that might occur concurrently with rotavirus may also be needed, as well as nursing care, such as rehydration of pigs with rotavirus that have diarrhea.

Salmonellosis

The disease salmonellosis results from infection with the bacterium *Salmonella*, which can be spread in a variety of ways. Pigs can contract the infection from other infected pigs or by coming into contact with infected feces. They can also contract *Salmonella* if their feed is contaminated. A United States Department of Agriculture, Animal and Plant Health Inspection Service (USDA/APHIS) study of feed samples from three hundred finisher operations showed that the incidence of *Salmonella* was less than 6 percent. Of the operations in which the primary grain was corn, 5 percent tested positive for *Salmonella*, and of those that fed some other grain besides corn, there was a 20 percent incidence of this pathogen.

There are different types of *Salmonella* "bugs," and they cause different diseases. Thought to be the most common in swine is the one known as *Salmonella choleraesuis*, which can lead to pneumonia, gut problems, and septicemia, which means the infection spreads throughout the body.

If the respiratory system is affected, swine may cough and have trouble breathing. If their gut is affected, they develop diarrhea, which can come and go and, of course, lead to dehydration. When salmonellosis becomes septic, other internal organs besides the lungs are affected, such as the liver and spleen. Other signs of the disease include loss of appetite, fever, and depression.

Swine with this disease need treatment with an antibiotic, and you'll need the help of your veterinarian to determine exactly which one will work. Salmonellosis has been linked with poor sanitation, overcrowding, and bringing in feeder pigs from multiple sources, so heed advice about keeping pigs clean and comfortable and buying stock from one source whenever possible.

Salt Poisoning

Pigs can develop salt poisoning by consuming excess salt in their diet, but more commonly the problem begins when they are deprived of water, then drink plentifully. With water deprivation, salts accumulate in the body, and when the animal drinks, water is drawn into the central nervous system. Swelling in the brain results.

The signs of salt poisoning include restlessness, constipation, thirst, and depression. Salt poisoning can lead to blindness, deafness, convulsions, and even death.

Hogs can be inadvertently deprived of water if their source freezes, if there is not enough space around the waterers for the number of animals you have, or if your automatic waterer stops working. If you have to give pigs medication in water, they may not like the taste and may stop drinking. Ensuring that your pigs have access to water is very

important, so if the taste of the medicated water is a problem, try adding flavored gelatin to mask the unpleasant taste.

Scouring

Scouring, or diarrhea, can be caused by a variety of factors, ranging from dietary problems to infectious disease. It can be lethal in a very short time, and unless you are absolutely sure of the cause, consult a veterinarian to determine that cause as well as a course of action. Prompt treatment will not only increase pig survival but also will help maintain reasonable growth in the face of a health problem. Pigs with scours usually become dehydrated and need rehydration. See Dehydration on page 241.

Shock

Shock occurs when there is inadequate blood flow to the body tissues. There are numerous causes, including severe stress, trauma, serious infections, and dehydration. I've found that trauma is the most common cause.

Swine in shock may be prostrate and have a rapid pulse, rapid breathing, and low body temperature. Hogs in shock need to be kept warm and dry and should receive treatment for dehydration.

Immediately administer an intraperitoneal injection of epinephrine if you suspect a hog is in shock. It usually works fast. The cause of shock also needs to be addressed — wounds must be treated, for example.

Splay Leg

Splay leg is a condition of piglets in which their legs are splayed out from the body. The defect will be readily apparent soon after birth. Hogs can have a genetic tendency toward this

WATCH THE SALT

Generally, the salt content in feed should be less than 1 percent.

Splayed legs are readily apparent.

condition, which underscores the importance of selecting hogs carefully for breeding.

Another cause of splay leg can be slippery floors, so provide good footing for your animals. They need to be roughened and even cleated where a hog is expected to climb. Too-smooth or too-wet surfaces can result in foot and leg injuries and splay-legged hogs and pigs. To reduce this problem hogs can be broken to dung and urinate in one corner of a pen: Wet that corner and transfer wastes there when you introduce pigs to the pen.

Stomach Worms

There are several types of internal parasites that occur in the stomachs of swine. Among them are the red stomach worm and the large roundworm.

The red stomach worm is likely to be seen in hogs raised on pasture. The only sign of this parasitic infection may be less than optimum weight gain, but infestation can also result in stomach irritation and ulcers.

The large roundworm is more common. Indeed, the females are prolific — they can lay over a million eggs per day. The eggs can be transported on your boots and are pretty resistant to disinfectants, although heat or sun will render them unviable.

Growing pigs are more often affected by the large roundworm than are adults. Also, as with the red stomach worm, infection is more

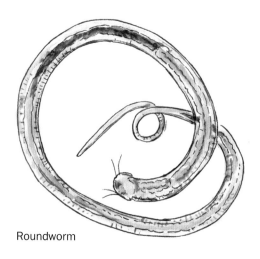

Roundworm

NEEDLE DISPOSAL

Safe disposal of needles is a concern, and in some areas there are regulations governing it. Generally, it is advised that the needle be broken off and inserted into the syringe before disposal, but it can be difficult to break off some of the heavier needles used for swine. We return needles and syringes to their rigid plastic cases and dispose of them by burning. Check with your veterinarian for further advice about how to best dispose of used needles.

common in swine raised on pasture. Although considered a stomach worm, the large round-worm migrates to the lungs, and that's where it can cause problems. Infested pigs may have trouble breathing and the asthmatic cough characterized by a "thumping" that I mentioned earlier.

The treatment for stomach worms is an effective worming program, as well as improved sanitation.

Swine Dysentery

Swine dysentery is an infectious disease that is usually seen in growing and finishing hogs. The infected feces of pigs, birds, rats, and mice are the source of the spread of the disease. Although most hogs recover, the death rate from dysentery can be as high as 30 percent.

When dysentery starts, you may see yellow- or gray-colored feces, then mucus and blood in the feces. Hogs with dysentery may or may not develop a fever or go off their feed.

Antibiotics administered in food and water generally are used to destroy the organism that causes swine dysentery. There is a vaccine available, but its results have been mixed. Good sanitation, such as providing hogs with clean housing and the elimination of any rodents, will help control the disease.

It can be hard to tell dysentery from other diseases that affect swine, so you'll need your veterinarian's help for diagnosis and treatment.

Swine Influenza

The hog version of the flu, swine influenza is usually caused by the Type A influenza virus. It is a disease that can be transmitted both among pigs and between pigs and humans. It can also be spread to pigs via lungworm eggs, which are passed in the feces of infected pigs, then eaten by earthworms, which are in turn eaten by the pigs.

The signs of this disease in pigs are similar to those you would have if you came down with the flu: It comes on suddenly and causes a deep, dry cough; fever; and loss of appetite. The animals may have a discharge from the eyes and nose. Although hogs usually recover from the flu, what often complicates the picture is a secondary bacterial infection. In addition, sows infected during pregnancy can have small, sickly litters.

There is no specific treatment for flu other than good nursing care. Pigs should be kept comfortable and under minimal stress. They might need antibiotics in their drinking water if they also develop a secondary bacterial infection.

Swine flu is often brought onto a farm when new animals — feeder pigs or breeding stock — are introduced. It tends to occur in the fall and winter months, and after there have been wide fluctuations in temperatures, which stress pigs. Some pigs carry the disease even

DISEASE OF THE MONTH?

So often I hear hog producers in a panic over the "disease of the month." More often than not, these tend to be diseases that are affecting commercial hog producers with large numbers of swine in confinement operations. This or that disease gets a lot of press, and pretty soon everyone is alarmed. These large commercial operations have a continuous flow of pigs in and out, which is more conducive to the spread of certain diseases than is the environment on small farms.

Of course, anything is possible, but if you select your animals for hardiness and manage your farm well, it's unlikely you'll have too much trouble. You can further protect your animals by keeping them away from perimeter fences and by prohibiting feed trucks anywhere near your hog lot; the undercarriage of a vehicle can carry disease organisms. In addition, the large ventilation fans in some hog houses have been found to distribute disease organisms to a distance of up to 12 miles. If you know there's an outbreak of something like TGE in your area, stay home to have your cup of coffee instead of having it in the town diner.

though they have no signs of illness. That's why it's so important to shop carefully for new pigs, make sure they're disease-free, and do everything else possible to prevent exposure of your hogs to anyone or anything that might introduce disease.

Transmissible Gastroenteritis (TGE)

Transmissible gastroenteritis is a serious, highly contagious disease caused by the coronavirus, and it's one of just a handful of swine ailments around today that deserves any press it gets. It can affect pigs of all ages, but young pigs are the ones that become most ill.

TGE tends to occur in winter months. Pigs get it when they ingest the virus, which they can get from many sources: infected pigs, fecal contamination of your boots as you walk into the hog pens, and even dogs and birds.

Little pigs with TGE experience vomiting and diarrhea, become dehydrated, and often

die, especially if they are only a couple of weeks old or younger. Older pigs also have vomiting and diarrhea, and they may go off their feed; sows might develop a fever. However, older pigs often recover from the disease.

A milder, chronic form of this disease also occurs that is thought to be due to more than one "bug."

There's no specific, effective treatment for the virus that causes acute TGE. Good nursing care is crucial for pigs with acute TGE; they should be treated for dehydration and kept in a warm, dry place. In the chronic form of the disease, other "bugs" may be involved, so antibiotic treatment may be indicated.

To avoid TGE in your herd, be careful not to buy or bring in infected animals. Don't go to other places where there are hogs, or hog producers, in the same clothing you wear around your hogs. And be sure to keep birds and dogs away from your herd.

Trichinellosis

Trichinellosis isn't nearly the problem people believe it to be — the fear really is needless. Trichinellosis in swine is a parasitic infection. It can be transmitted via undercooked meat to humans, in whom the disease is called trichinosis.

Improved swine management and increased public awareness of the importance of thoroughly cooking meat have effectively eliminated trichinosis. In fact, the last I heard about the disease in people, it was associated with eating bear meat, not pork.

Whipworms

If you have pigs from about 2 to 6 months of age that develop bloody diarrhea with mucus in it, suspect whipworms. Pigs that ingest whipworm eggs in feces also become infected.

An infection with whipworms, also known as *Trichuris suis*, leads to dehydration and wasting. Pigs with a whipworm infection may also go off their feed and become anemic. They can die.

As with most types of parasitic infections that affect swine, good sanitation and an effective parasite-management program will control this problem.

TREATING WOUNDS

In hogs, the most common wounds are scrapes, cuts, and punctures. They tend to be shallow and will often heal with little or even no treatment. An aerosol wound spray, such as Blue Lotion, or tamed iodine can be applied quickly, disperses over the wound site, dries rapidly, and can be used without restraining the hog.

Some hogs become skittish when they hear the hissing sound of an aerosol. Fortunately, many wound sprays are now available in pump and squeeze bottles. Our veterinarian even packages a treatment he prepares for wounds and to treat navels into squirt bottles, which are available in several sizes.

With very deep puncture wounds, however, some products may not penetrate fully. You run the risk that the wound will heal over at the surface but leave an infection underneath. For such wounds, obtain from your veterinarian and keep on hand a prescription product made to combat infection; it's more potent than what you'll buy over the counter. When using any spray around a hog's head, be careful not to get the product into the eyes or ear canals. Watch out for your own comfort, too. Tamed iodine is supposed to be milder than full-strength iodine, but it still always seems to find its way into that one little nick you have on your hand; you'll find that the word "tamed" is a subjective one!

11

MANAGING YOUR HOG OPERATION

Whether your swine venture numbers two head, two hundred head, or even two thousand head, it is very much a business and should be operated as such.

To be successful with hogs you have to do two things: get muddy and get serious about making a profit. And to achieve the latter you have to know where every hog in the herd is, how it is performing, and how much it is costing you.

Record Keeping

Experience teaches that the best-looking and best-doing animals in the herd are seldom, if ever, the same ones. The only way to find those best-performing individuals lies in maintaining detailed records on herd performance.

We once owned a Hampshire sow that we used to produce F_1 Duroc × Hampshire boars and gilts. Her first litter was born early in a hot August, and with her pigs at a bit older than 3 weeks in age, she went off her feed and began to suffer heat stress. For nearly 48 hours she didn't eat — but she did continue to nurse her pigs. The toll it took upon her was visible all her life.

At a time when she needed food to make her required growth as a first-litter gilt and nurse her pigs, she suffered a serious setback, and for the rest of her life she remained below average in size. In the breeding pen and the gestation lot, she was always the small sow that didn't fit and didn't look right.

The record book, however, showed that it was actually the other sows that lagged behind her. Across 10 litters, she had an 11.2-pig weaning average and also averaged more than three keeper boars per litter. It is key on a small farm that every creature pay its own way. This can only be determined with careful record keeping and making animal selections based on their records.

Field Notes

Perhaps the place to begin a discussion of swine records is in the farrowing house. Here two great opponents meet: the hog farmers who believe they are too busy to put things down on paper and those who are drowning in a sea of data, yet thirsting for more numbers to crunch.

Nailed above every crate gate in our old eight-crate pull-together was one of those little

---

The best tool for guiding a swine operation is a set of detailed and regularly maintained business records.

clothespin-type paper clips. Snapped into each one was a 3 × 5-inch index card. Onto that card went the sow's ID number, the boar to which she was bred, her farrowing date, her litter size, the pigs' birth weights, and our comments on the sow's performance.

We have seldom in 30-plus years of hog raising operated with more than 20 sows, but even with a sow herd you can number on one hand there can be too much data to keep just in your head. Those index cards might have wound up fly specked, written on in three different colors of ink, and dog-eared, but they were easily at hand for quick notation where the action was really happening.

When the sows were pulled from the farrowing quarters, the note cards were taken to the house and deposited in a file box for future reference. For an investment of less than $2, we were able to build a performance record on our most important asset, the sow herd.

Another couple of dollars buys you a little notebook called a "shirt pocket handbook." Available from purebred-swine groups and many mail-order farm-supply houses, it contains a gestation table and a log page for each litter farrowed. It has spaces to record sire and dam, farrowing date, and more, plus a space to record each pig born by its litter and individual notch.

It thus becomes the place to begin amassing information on your individual pigs. You can log in such data as teat counts, health treatments given, and comments on each pig's merits. In the pen and on paper, the pigs will begin emerging as individuals.

Permanent Records

The shirt-pocket file can be backed up by a desk model into which the data is transferred to become part of your operation's permanent records. Into this desk notebook or computer file go the field notes and end-of-the-day observations on everything from pen assignments to ration changes.

Dad paired his shirt-pocket notebook with one of those big wall calendars that banks and elevators used to give away at Christmastime. There was a large square around each date with room to write all sorts of things. Over morning coffee, Dad would transfer the previous day's notes onto the calendar, and by the end of December he had a yearlong log that could be rolled up tightly with his notebooks, closed with a rubber band, and stored neatly in a drawer for future reference.

Today, those notes are just as apt to be keyed into a computer, but the point is that a permanent record system is important. As I write this, my permanent record is less than an arm's reach away from my favorite chair. It helps me keep track of the little points that can make enterprise management so very much easier.

But I haven't always been so on top of it; there have been times that lack of organization got the best of us. For example, while at the 25-sow level, we operated for a time with two herd boars. With 25 all-red sows and two all-red boars, the stage was certainly set for at least some confusion, and we lost track of the service sire of a couple of litters due to poor record keeping. Some pretty good purebred pigs had to go into the butcher pen simply because things were left to memory or written down on scraps of paper torn from feed sacks that ended up in the washing machine instead of the record book.

Record-Based Management Decisions

Filing away all of this data in a neat manner until tax time is really just a small part of operating a swine enterprise with good records. You also have to rely on them to guide you when it comes time to make tough choices.

Here's one scenario you might come up against: With 15 to 20 sows, the butcher-hog market going south at $1 and more a day, and corn pushing through $3 a bushel, you have to face up to some rather hard decisions. How deep to cut? Which animals to cut? Which areas to cut back in? Where to spend money to do the most good? Yes, sometimes the best thing to do during troubling economic seasons is spend some money.

I often make a statement that gets the juices flowing in just about any group of farmers, and I will stand by it here, too: There is not a herd or flock of any livestock species anywhere that would not be made better by simply removing the bottom third of the animals it contains.

Sometimes the "my herd is bigger than your herd" pride gets in the way of good management decisions. But your owning and operating costs would be trimmed dramatically by removing animals on the low rung, and your income would not be trimmed nearly as much as you might think. The poor performers always rob from the attention and resources that should be directed to the better performers in any herd or flock.

How to know which animals to remove is the challenge. Just strolling through the herd and choosing strictly on visual appraisal can give disastrous results. The thin sow with the razor-sharp backbone would look like a sure candidate for culling, but she may have just weaned a litter of 10 big pigs and done as well with her previous three litters.

The next sow you encounter may stick in your mind for having weaned nine pigs, but what if three of them weighed far less than the herd average for weaning weights? Next up are two females that have just weaned eight-pig litters, but for one it was her first litter; for the other, her eighth.

Only by having good records on current and past performance can you make the hard decisions. You can't trust every litter farrowed to memory, let alone every pig, and if you rely only on sight appraisals, you might just wind up with a lot of slick-haired sows that farrow a three-pig litter once every 12 months.

Sow-Herd Performance Records

Nationally, the average sow herd turns, or replaces, 40 percent of its numbers yearly, and many of the commercial gilt suppliers expect a 40 percent culling rate of the females they supply before even the second parity.

As a sow ages, her farrowing times tend to grow longer, and there is an accompanying increase in the number of stillbirths in each litter. You will see the normal wear and tear of aging, too, including some unproductive udder segments. Over time some sows will also succumb to some sort of injury or malady that impairs their reproductive performance.

I will generally give a gilt that farrows a litter of six or seven a second chance (a couple of them have turned into sows that regularly farrow litters of seven). A sow with a small litter will also get a second chance to redeem herself if the problem was obviously due to management or environment. For gilts, a good number for both litter and weaning size is eight; for sows, it is nine. Still, it is best to acknowledge a problem performer in the early going, get her out of the herd, and bring in a younger replacement with better prospects. A few years ago, we rebuilt our Duroc herd and went through nearly a dozen females before finding the one female to build on.

Records of age, litter size and numbers of pigs weaned, number of litters, injuries, birth defects, farrowing difficulties, and environmental factors that may have had a role in productiveness or lack thereof all help me to assess the sow's or gilt's performance and to decide whether to keep the old (or young) girl or send her down the road.

Financial Records

Good financial records make it possible not only to determine your production costs but also to pinpoint weak areas within your enterprise. Repeated studies of producer records across the country show that often the only difference between the most profitable producers and those just scraping by is how well they do at that most basic of all management practices — purchase for the swine enterprise.

An old but valued tool for tracking and recording income and expenses is the multiple-column record book. Available for a couple of dollars from most stationery stores, it provides spaces for recording the date of the purchase, what was purchased, from whom, the amount of it, and the cost; additional spaces allow you to break it down by specific category (grain, protein supplement, veterinary, etc.). This data can be recorded in just a few minutes at the end of the day or when bills are paid.

Much of this is information already carried in the farm checkbook. By breaking it down in a columnar fashion, though, you can backtrack through your enterprise in a number of ways that will help you assess its health and success. Records, such as a computer spreadsheet, can give a good overview of everything, from what it costs you monthly to stay in business to gradual creep in the costs for very specific inputs. On most small farms, the primary expense

AVERAGE LENGTH OF SERVICE

We had one sow that was with us for seven years and 12 litters. But most sows' usefulness lasts three or four years on a small farm, less in a large commercial operation. Most boars are gone by 30 to 36 months of age, because they have grown so large — often to more than 500 pounds — and because by then their daughters are coming into the herd, which means there is a pressing need for a new unrelated boar in the one-male herds we have been focusing on here.

to produce pork is feed costs, and the largest expense there is for grain. On some of the larger farms and pork factories, interest and energy costs are beginning to rival feed costs.

If costs are beginning to creep up unduly in a specific category, they will jump out at you if you use this type of record keeping. Records can also help you target areas in which cash savings may be possible. At this writing, for example, the elevator east of our hometown is charging 20 cents more per bushel for yellow corn than the elevator west of town. Price columns that don't move when they should denote the need for some comparison shopping for a particular input.

With a few months of such records in the book, you will be able to compute costs to maintain a brood sow, produce a 40-pound feeder pig, or crank out 100 pounds of market hog. Those figures are key to assessing the financial well-being of our venture.

Analyzing Recorded Costs

If costs to produce are exceeding available selling prices, your venture is indeed ill-fated. And if all costs to support said enterprise are not directly traceable to it, it may actually be subsidized by another kindred or connected venture. If your hogs, chickens, and cattle are all being fed out of the same reserves of corn and you have no way to ascertain how much of that grain is going where, you indeed run that risk.

There are two sets of costs involved in swine raising: fixed costs and variable expenses. Fixed costs are the costs of owning hogs that have to be paid year in and year out. They include interest and payments on principal, taxes, insurance, building costs, and the like. Variable costs are those that the producer

> ### STUDY RECORDS MONTHLY
>
> Study your records at least once a month for trends. Too many producers look on records as something to work at to please the tax man rather than something they can use to improve their enterprise.

can do something about and include feed, veterinary bills, energy, transportation, maintenance, and supplies.

Smaller-scale producers often pay too much for basic supplies and services. There is not much to be had in the way of discounts when purchasing feed just 200 or 300 pounds at a time (though some feed dealerships do offer 1 to 3 percent discounts for cash purchases). What is interesting to note is that the supposed volume discounts offered to many large-scale producers really do not exist. The industry has been hard pressed to demonstrate any buying advantage for producers even to the thousand-sow level. Real advantages exist only for producers large enough to totally bypass an entire group of suppliers.

Several years ago, I worked in livestock feed sales, and seldom did we have more than a few-dollars-per-ton leeway for dealing with our largest customers over our smallest. Our very best prices came as a result of such things as crop increases and competition among suppliers. The big operators did seem to take greater advantage of such specials when we had them available, however. They had the storage space and perhaps the greater liquidity to move on such bargains when they did occur.

I can recall one summer when large acreages of soft wheat in Missouri and Illinois sprouted problems with wild onions. The wheat harvested from those fields was so laced with the onions that it could only be used in livestock feeds. Some 12 percent complete swine-finishing rations actually slipped to considerably less than $90 per ton whether you bought one or one hundred tons. A place to store it and the money to buy it were all it took to take advantage of such an opportunity.

On feeds that are consumed in small amounts, regardless of operation size, there are few opportunities for volume-purchase discounts. Even in the Midwest, some elevators order and stock swine feeds in volumes as small as 500 pounds.

A good, long-term set of records can help to point up things like seasonal buying advantages for various inputs. Sometimes, for example, the best time to buy straw for bedding is at the end of winter, when demand is down and sellers are clearing storage space for the new crop of straw coming that summer. However, don't be fooled by every cliché. At harvest is often the best time to buy a supply of grain, but in years of really short crops, the best buys may not come until the following summer.

MONITOR VARIABLE PRICES

The time to take a look at variable prices is not when the increases are coming fast enough and in big enough increments as to be painfully obvious. Instead, scan your records for emerging problem patterns at least once monthly.

Recording Sales

Along with inputs, you should keep equally detailed records on your sales. The columnar notebook allows space for recording the date of the sale, type of sale, to whom the sale was made, selling price, and amounts sold by category (butcher stock, feeders, breeding stock, cull animals).

Contrary to what many believe, the advent of high-volume swine producers has not eliminated the traditional seasonal aspects of swine production and their consequences. Feeder pigs still bring the highest prices in March and April, because few producers want the difficult task of farrowing in the cold winter months. Their prices are most depressed in early summer and late fall, following the spring and fall farrowing patterns that are preferred.

There is a general downturn in red-meat prices in November, when all of those spring-farrowed pigs are coming to town and the big holiday meals build on a turkey. June, July, and August often bring the highest butcher-hog prices of the year, because the demand is up for barbecue meats while the pork supply is down when there are fewer cold-weather-farrowed pigs.

Knowing who bought what and how well they paid helps you build reliable markets and know where to target the various segments of your production. Does volume buyer John Smith readily pay market top for 60-pound feeder pigs but hedge on 40-pounders? Why? Didn't the ABC buying station pay more for cull sows than the XYZ order buyer?

And then there are the kickers in the deck. Are enough sales falling in the month the farm payment is due? Are you dumping cull stock in late fall when all red-meat prices trend downward? Are hogs fed in one season of the year

TEN CRISIS-MANAGEMENT STEPS

In a situation where plunging prices or drought conditions force herd reductions, there are steps you can take to make an orderly retreat that preserves your herd as a viable unit:

1. Recognize the emerging problem as soon as possible, so you can take steps before salvage values plummet.

2. Cull older sows first whenever possible. They represent the most dated and what should be the least productive genetics in the herd. The age of "older" sows is usually about 3 years and older.

3. Females in their first three parities must be considered the future of the herd: They have the longest productive lives before them.

4. Apply the best remaining feedstuffs where they will do the most good: to females in late gestation, nursing females, and young pigs.

5. Consider alternative feedstuffs, which may be cheaper at times (e.g., replace a portion of the corn in a ration with grain sorghum, or use lysine to replace a portion of the soybean oil meal).

6. Adjust weaning times to best utilize available feed supplies. (When necessary, I've dropped weaning to 28 days.)

7. Buy feedstuffs only as needed.

8. Sell the pigs as feeder stock.

9. Use the depressed prices to upgrade your breeding herd by buying herd replacements on a down market.

10. Pat yourself on the back for managing the situation.

Grain sorghum

taking substantially longer to get to market than those sold in another?

The most interesting reading on the small farm should be the record books. They give the look beneath the hood that is vital in assessing how well that farm is doing and where it is heading.

Marketing Hogs

Marketing is the neglected chore on nearly every hog farm. Producers approach it with dread, with little knowledge of what drives their markets, and with a belief that no matter what they do the market will always be stacked against them.

Many years ago, local radio stations reporting the hog and cattle markets from the National Stockyards in East St. Louis, Illinois — just across the river from St. Louis — would give the names of the sellers at the market that day who had the higher-selling animals. It was quite a feather in the cap down at the coffee shop to have landed the high price of the day and be mentioned on the radio.

At first glance, setting the market-topping price would seem a very valid goal, but it can be one with a painful backlash. Producers who hold out for the very last penny that the market has to offer often find the market buckling rather than following their expectations. A more reasonable goal would be to make all your sales within the top 10 percent of market prices for the year.

The basic cull-stock, butcher-hog, and feeder-pig markets function as wholesale markets. They are set up to move a lot of volume, and all too often the only premium they have to offer is for maintaining that volume. Buying stations across the Midwest often hold an extra 50 to 75 cents per hundredweight under the counter to draw those numbers when they are needed. And they favor those who can provide the numbers with that premium.

Quality Counts

Still, in recent years the market has begun to at least pay lip service to questions of carcass yield and quality. In the early 1990s, there were many rumors of buying stations and order buyers turning away butcher hogs because they just didn't measure up. Raisers were told their butchers were too lean, lacking substance, and prone to produce poor-quality pork. A lot of small farmers were nipped by this, and many overreacted to the point of discontinuing a basically sound swine enterprise or backing away from starting one. They had fallen into the trap of thinking that they were just too small to compete anymore, too small to pay for quality inputs, and too small to rise to any challenge. It was a costly mistake — but one with a fairly simple and inexpensive solution.

The answer to the quality challenge is often just one good boar away. During the hard going in the early 1980s, I sat ringside at more than one auction where that kind of boar — of good quality but a less popular breed — sold for as little as $200.

But the swine industry wasn't built on $100,000 buildings and $25,000 pickups — it's the hogs, stupid. A valued banker friend once told me that a sure sign of a hog raiser on his way to trouble was a request for a loan for a big, flashy farrowing house. In fact, dollars in any shape or form other than at the bottom line at the end of the year are a pitifully poor way of scoring a venture. A few years ago, a Duroc boar made lots of waves with a $35,000 selling price, but just 18 months later I bought one of his sons for only $125.

There are boars out there that can shave off ¼ inch of backfat from their offspring, or add 1 square inch to the average loineye area. Some of them are like pepper on a good steak — you just need to use a dash — but the only way to make hogs truly better is to breed them to better hogs. It matters not a whit where they were born, what they ate out of, or what kind of truck they rode to town on — it is the blood and breeding that tell the tale.

Study the Market

Doing a better job of marketing often means acquiring more knowledge about how your available market outlets operate. For example, at times we have taken one or two large boars to market after using them in the breeding herd to produce one, two, or three pig crops.

The fate of most such boars is pepperoni and other highly spiced sausage products, but not every market outlet has access to buyers for such specialized production. Often, mature boars are held at a local assembly point, enduring much stress and shrinkage, and then are

BUTCHER-HOG MARKETING APPEAL

Ten good butcher hogs will draw a good dollar just as often as a hundred in an area with competitive market outlets. A small group of butchers can create added appeal in several ways.

- **Have pigs sorted closely, and they should be uniform in both type and weight.** In a small group, a pig out of the loop by even as little as 15 pounds will show up vividly. Information from a national survey conducted by the USDA Animal and Plant Health Inspection Service shows that product uniformity is very important in the pork industry. Packers often give premiums to producers who bring uniformly sized pigs of similarly lean quality for slaughter.

- **Be a market watcher.** Ten head are easier to load than a hundred and can get to market quickly on days with low runs or when prices take a sudden jump.

- **Don't back away from incentive programs.** Too many small producers respond negatively to programs such as grade and yield buying, which pay a rate based on carcass merit for quality hogs. These small producers have an unfounded fear that their hogs can't cut it on the packinghouse "rail," where hogs are paid for on the weight and merit of the dressed carcass. The producers lack confidence and feel they can't or shouldn't compete. They use small size as an excuse to not try harder. Actually, on the rail is the fairest place to compete. There are no judge's opinions there; every hog competes straight out head to head.

- **Don't try to hide a dog in a bunch of good hogs.** Instead of trying to slip a poor hog past a buyer, acknowledge it as a mistake you can and should eat or at least send along to the sale barn.

- **Know your market people.** Listen to your buyers' wants and needs, and as much as possible, try to produce the hogs they like to buy.

Note: The truest view of hog quality emerges when they are slaughtered, dressed, and awaiting processing: fat cover and lean yield can be measured exactly. The hogs are suspended from the packinghouse rail and the carcasses fully exposed for quality assessment.

Many packers now want to wait to pay for butcher hogs until they see them on the rail, then offer fairly modest premiums for those that grade and yield better than the average for hogs of the same weight. The greater the red-meat yield, the more valuable the hog is to the packer and the greater the premium payment to the producer.

The poorer-quality pig will generally bring as much as it can at the market of last resort — the sale barn. There are, however, buyers who specialize in odd lots and misfits and may risk trying to increase such a pig's weight or quality through additional care and feeding.

moved along to a regional market. I found one local buying station that bunched such boars for shipment to an East Coast processor and paid as much as $2 to $3 more per hundredweight for them than other available markets.

An extra $10 to $12 per head for a couple of animals sold each year may not seem like much, but the first boar this processor bought paid for the phone calls to find which market was paying the most for heavy boars, and that knowledge became a little bit of a perk that I was able to pass on to the buyers of our young boars. In a year or so, they would have similar boars to sell and, hopefully, would remember who gave them the tip about this market and return to us for new boars.

A bit of study of your local scene can reveal the many things that make local markets rattle and hum:

- At most markets, highest prices come in the first four days of the workweek. Mondays and Tuesdays are generally the strongest days.
- Prices generally trend lower on Fridays, since the buyer usually has to carry the hogs through the weekend before processing them. The same is true of hogs sold on the eve of a holiday.

- Some markets have a broader definition of handy weights than others. Some will be more receptive to hogs less than 230 pounds or more than 260 pounds.
- Some markets eagerly seek culled breeding stock, others buy them as a courtesy or a cost of doing business, and others won't handle them at all.

The prices each market offers for culled breeding stock reflect its interest in this market segment, which is quite substantial. One area buying station won't buy them but will ship them to a larger, regional market if space is available on their trucks. One neighbor sent a big boar this way and, due to shrinkage and fighting stress, suffered nearly a $100 loss in the boar's value in just a few hours.

There are market outlets that do reward loyal patrons and not just with the checks they write. We once sent 10 thin sows to an area buying station at which the operator directed the trucker to take them to a weekly auction a couple of miles farther down the road. There they brought 12 cents a pound more than the station buyer was authorized to pay — from a local farmer, who bought them to glean his cornfields.

TRAVELING HOGS

Most finished hogs — more than 55 percent of them — are sold directly to the slaughter plant. Only about 11 percent of U.S. hog businesses sell for slaughter through auctions, according to the USDA Animal and Plant Health Inspection Service.

Hogs also travel a distance to market, although this varies with the region. More than 80 percent of hogs travel 200 miles or less to slaughter, and more than half travel 100 miles or less. They travel farther in the Southeast, but in midwestern and north-central parts of the nation, about one-third are within 50 miles of a slaughter plant.

Small Specialty Markets

Much is made of what is now called "niche marketing," but selling to rather narrow, very specialized markets has been a selling option small farmers have tapped into for generations. Some of them are rather short-term—if not one-time—deals, but others have grown into family businesses with the potential to thrive for decades.

Here's an example of a very short-term market option: I once had two little Spotted pigs go into a young girl's Easter basket and received a $10-per-head premium for them.

As an example of a marketing option that persists, I recall a story about a modest-sized purebred producer. He began supplying roasting pigs and even doing the roasting for parties and celebrations as a means of marketing hogs that were not of the quality needed for breeding stock. This market grew to a level of two or three such events each week and soon replaced purebreds as his primary swine venture.

Selling Show Pigs

A rapidly growing specialty market now seen in many parts of the country is the raising and sale of show pigs. These are pigs in the 35-to-60-pound weight range and of such quality as to be considered good candidates for the show-ring. With a good many families moving to a few acres in the country and becoming part-time farmers, a growing number of youngsters are seeking project animals but do not have the time or space to give to a traditional sow-and-litter project.

At some county fairs, the market-hog show classes have totally replaced the breeding-stock shows. Some producers are even organizing into marketing groups to sponsor once- or twice-a-year auctions of such pigs. As many

SHOWCASE YOUR PIGS

When you sell quality stock at a show-pig auction, you have a public showcase that can lead to further sales back home.

as 10 producers may come together, consign 10 head of pigs or so each, hire an auctioneer, select a sale site, and pay for advertising collectively. A number of pig producers may even join with an equal number of club lamb producers to sponsor such a sale. Some purebred producers are even forgoing the traditional fall sale to offer up some of their best spring pigs in a March or April show-pig auction.

The show-pig market is seasonal, with pigs of the right age for the summer and fall fairs most in demand. This generally means January-, February-, and March-farrowed pigs for sale in March, April, and May.

Pigs sold for show are the cream of the crop, so early on sales might be small. Among the most active in this market are the small and midsized producers with some previous experience in fitting and showing hogs.

Barrow pigs need to have been castrated early and fully healed. While not exactly in a fitted state, shoats are presented in a condition over and above that normally seen in barn-run feeder pigs. They are absolutely free from parasites and are kept well housed and bedded to maintain quality haircoats. They are the big pigs in the crop, display exceptional muscling, and are free from structural faults. They may be purebred or F_1 animals that just missed the cut as breeding animals.

Show pigs are not from thrown-together matings. They reveal a lot of black (Hampshire,

Berkshire, or Black Poland) and red (Duroc) breeding, to take advantage of the growth and muscling strengths of those breeds. Their numbers are limited at an auction, to maintain quality and to prevent a price break from occurring in such a public setting.

To further increase demand, some producer groups even sponsor their own jackpotted market-hog shows. They schedule a show with a date that will catch the pigs they have to sell, offer a top prize (sometimes the only prize) of between $250 and $1,000, and open the show only to pigs sold by group members. This show might be scheduled for a week or two ahead of another event such as the county fair to give youthful patrons a bit of added experience. Other groups may offer a $50 or $100 cash award and a trophy to any pigs from their offering that top various area shows and fairs.

SHOW-PIG PRODUCERS: ON THE CUTTING EDGE

Show-pig sellers are also offering a bit of expertise. Pigs selected for sale should be currently popular types. The producer makes him- or herself available to the young buyers to answer questions about fitting and showing, and the producers attend the various shows to lend support to their buyers.

Show-pig producers are a bit on the cutting edge. They tend to be knowledgeable about bloodlines, are breeding current types, and are familiar with the show-ring scene.

Pork Sales

For some years now we have raised some of our Willow Valley hogs for the pork trade at the farmers' market. A local abattoir uses our hogs to produce a very lean, whole-hog pork sausage that we sell by the pound at our local farmers' market. It is a USDA-inspected and -approved facility that processes, wraps, and fast-freezes the product for us.

We're no threat to Bob Evans or Jimmy Dean, but we do wind up with a premium product that we're able to sell for $2 a pound or better. We have the pork ground 80 percent lean to 20 percent fat (quite a bit leaner than most store-bought varieties) and only mildly seasoned. Spices can be added to sausage right up to the time it is eaten, but removing them to suit someone's taste is quite another matter.

The slaughterhouse freezes the sausage in 1-pound sticks, and we pick it up and transport it to the market in simple ice chests. These markets might be the local farmers' market or doorstep sales. I know of one farmer who sells sausage products from a freezer carried in the back of his pickup and plugged in whenever the truck is parked at a market site. At least in Missouri, you do not need a special license to do this as long as the meat comes from a USDA-regulated facility. We use the local processor's packing label, but our own label with details about the lean content and additive-free nature of our sausage would, no doubt, help sales. Alas, such custom-made packaging is very expensive and has to be bought in considerable volume.

There are some of us who share a fairly common sentiment that family farmers should be freer to directly process and sell their own production. They can't, however, because the needed inspection services are not in place. I think a farmer should be allowed to process and direct-market the pork output of up to five hundred head of hogs per year. That would be just a thousand cured hams and bacons to sell each year — hardly a threat to even the smallest of supermarkets.

In a direct-selling situation, you as producer have even more responsibility for quality control than if you were answerable to a whole phalanx of tax-and-license-fee-financed inspectors. You have to look at the end consumers of your product and know that your success comes from their satisfaction and that alone. Two quality cured hams will now often bring as much as or more than an entire butcher hog on the hoof — but family farmers are now largely cut off from this market, which was once their traditional domain, through a series of nitpicking rules that do not always prove effective in guarding public health or food quality.

Sadly, the niche marketplace for pork is now almost exclusively local. Sales over a state line can't happen until some exemptions are made.

Pork sausage is just one pork product that has proven to be a good venture for small hog farmers.

CUSTOM FEEDING HOGS

Those farmers who prefer a less involved approach to the direct marketing of their butcher hogs may offer whole or half carcasses delivered on the hoof to a processor that has been mutually agreed on by the buyer and seller. I have seen these hogs advertised as "fed to order" or "custom pork product." The buyer can request a specific slaughter weight, additive-free rations, range raising, and the like, and the producer prices the hog accordingly. A substantial deposit is in order before undertaking to custom-feed a market hog. It is certainly a practice in keeping with the trend toward community-supported market gardening and subscription fruit growing.

A listing under a "Good Things to Eat" heading in the want ads of a nearby metropolitan paper may be among the best places to locate such buyers until word of mouth (the very best advertising) can build. Simple flyers pinned up in places such as health and specialty food stores and grocery cooperatives may also draw potential buyers.

Marketing Tools

I have used the weekly newspapers in the counties adjoining ours to offer boars, gilts, and feeder pigs with good results. Calls to the swap-shop programs on local radio stations have also paid off with sales for me. I put up flyers at local sale barns, too. You could try passing out business cards at busy places such as feed stores, elevators, or auctions when you hear someone expressing an interest in acquiring hogs.

We maintain a list of all of our past buyers, including their addresses and phone numbers. When a set of young boars is ready to sell, I can check that list for those who should be due for a new set of boars. Then I send them a postcard with a listing of what I have to sell.

One midsized breeder I often work with mails out his own newsletter, which features brief items and pictures of his buyers and lists the animals he currently has for sale. When his sales slow, he offers discounts on the purchase of two or more boars or a sale on older boars; once, he even included a $25-off coupon good on the purchase of one boar. This breeder uses the postal service, but if he had Internet access, an e-mailed newsletter would be less expensive and could display his offerings on the computer screens of many customers with a single touch of the keyboard.

A most imaginative selling tool he uses makes many buyers feel more at ease with selecting boars from him while also steering them to his higher-priced animals. The pigs he deems best in a group receive purple ear tags, and the second level gets blue tags. Buyers see this as a way of reinforcing their choices. The colors were chosen for a purpose. The very best hog in a show receives the purple ribbon, and it is awarded out of the top group, the blue-ribbon winners.

On another farm, if 10 gilts are bought at the regular price, an 11th is available for just $100. Years ago, one legendary breeder would actually give a free boar with every 20 gilts purchased.

We, too, give discounts to buyers — $10 to $25 off per head for purchases of two or more head and a similar discount for repeat customers. This generally works well, but at 8- to 16-month intervals, I have a set of brothers who take 10 to 12 boars at a time. They expect a bit of preferential treatment, and they get it, but too much of a discount and it's like giving them a boar for free.

The Hog Industry: Today and Tomorrow

Earlier, I pointed out the risk to small farmers of thinking they are disadvantaged or unable to compete because of the small size of their ventures. Quite simply, mind-set has been either the undoing of or the key to success for more smallholders than anything else by far.

It is far easier to fund and establish 10 different small farm ventures netting $2,000 each yearly than to create just one enterprise with a capacity for $20,000 in net earnings. On the small farm, at least a couple of those modest ventures can be swine based. We raise breeding stock, sell feeder pigs, and market pork sausage, all from a quite small group of sows that fit into a number of other ventures on our small acreage, including raising rabbits and chickens, hatching eggs, and writing about raising livestock.

We putter along with a sow herd that is too small to show up in surveys, use a lot of 20-year-old equipment (pile it all up, light a match, and for $1,500 I'm back in business tomorrow), and own nary a pickup or tractor.

MORE MARKETING IDEAS

It makes very little sense to work hard at all the other phases of swine production only to falter and fail at the last and most important task: marketing. Imagination and a dedication to quality will stand the small producer in far better stead than simply playing the numbers game.

Many are uncomfortable with the practice, but you must promote your pork enterprise at every opportunity. Pork may be promoted nationally as "the other white meat," but the only one who will promote your hogs and your pork is you. Toward that end, keep these points in mind:

- **You already have a market.** In friends, neighbors, family, and business associates, you have a ready-made list of potential buyers.

- **Opportunities to promote come in odd ways.** I know one pork producer who brings a new and novel pork dish to every potluck dinner his family attends.

- **You need to let people know where you are.** Get your name on the mailbox, and get a sign out in the road pasture. If they don't know you're there, they can't buy what you have to sell.

- **Group efforts can pay off.** Whether you're establishing a farmers' market, starting a sale, or just getting enough hogs together to fill a big truck, cooperative efforts can be an excellent way to get your name and services out there and on the tongues of prospective buyers.

- **Many folks shop on the Web these days.** You can present your website "store" as artfully as you like and invite people into separate parts of the "barn" or "range," where they can learn about your diverse ventures, prices, and purchasing logistics. You can include site photos of your boars and sows, descriptions of your raising practices, breed histories, taste-test results, recipes, tips, customer-satisfaction quotes, contact information, and links to breed associations and other informational websites. Someone in your neighborhood knows how to build a website, and paying him or her to do that for you if you are not computer savvy is an investment that will pay off in spades.

- **Good publicity helps your business grow.** Create your own press releases or seek help from those with whom you interact, such as breed association staff. We once bought the top-selling boar at a breeder's auction and got a nice mention in the breed association magazine, a couple of state farm magazines, and a couple of local newspapers.

- **It helps to be seen in the right places.** We attend Pork Day activities in the next county, take in nearby sales, attend hog shows, work with 4-H youngsters, and buy from our fellow small producers.

Still, we produce a quality product that breeds on for our customers. We're putting money in the bank, and we're content with what we're doing. And never underestimate the importance of that last point.

Growth: At What Price?

Much, too much, is made of growth in production agriculture. As I watch those about me pushing to add ever-higher numbers and build ever-bigger herds, I'm reminded of two things. The first is a line my grandmother often repeated: "No bird ever flies so high but what it doesn't have to light." The other gets even more to the point — growth for growth's sake is nothing more than the philosophy of the cancer cell.

There is absolutely nothing wrong with staying small, as long as you are committed to staying good. At times you may feel that you're striking out on a lonely and uncertain course, but such isn't really the case at all.

The average sow herd in the United States still numbers less than 40 head. Pigs are on farms from Hawaii to Delaware in numbers great and small. Just one sow producing 18 pigs in a year translates into more than 1½ tons of pork sausage. If you have a market for that much sausage at $2 per pound, you have gross yearly sales in the $6,000 range.

You do, that is, if your pork is of good quality, an approved processor is available, and you are comfortable with making direct sales and extracting that high level of production from a sow. At this writing, the small-scale pork producer must function as something of a maverick — a re-pioneer.

Before I am accused of coining a totally new word, what I mean is that smaller pork producers are in a position to lead an entire industry back to the point where the animals are humanely reared, the environment is well tended, the consumer is better served, and all hogs and producers are respected as worthwhile, contributing members of rural society.

Public support of high-volume production agriculture has begun to slip markedly. The quality of its output is being questioned and challenged as never before. The only farming sectors still in total favor with the consuming public and whose output garners a willingly given premium from that public are the organic growers and those producers who can be clearly identified as family farmers.

Corporate Competition

Conversely, from the farmer's perspective there is often doubt whether the small producer can survive, let alone continue to compete. Right now it is hard to ignore the big corporate producers — they are the 800-pound gorillas — but their warts are beginning to show. Close-packed swine populations are more disease prone, they are creating great seas of wastes, they sell entirely in a whole-sale market, they have fixed operating costs that are often as high as the variable costs on some small farms, they have odor problems that make them the classic not-in-my-backyard (NIMBY) enterprise, and many consumers are organizing to resist their existence.

There is some strong evidence that the big operations are seeing the writing on the wall and may be planning to get out of the pork business sometime within the next 20 years. Many of the big buildings that have gone up have done so with a 10- to 15-year cost-recovery plan. These operations are often just single arms of multinational corporations that can shed them as easily as some lizards drop their tails.

If you don't think this is the case, just look at the poultry industry's handling of broilers: incubators could be turned off, hatching eggs dumped, and workers cut loose, and the farms would still chug along just fine with all sorts of other income. On the other hand, few contracts for swine finishing extend for more than a year, and one of the largest corporate swine operations in the Midwest applied for bankruptcy protection a scant five years into its run.

Animal rights issues, environmental measures, and consumer concerns can and may all grow to a point where the pressure they will bring to bear on the large operations will be just too great. Some small farmers are already tapping into this resistance by pursuing niche markets for additive-free pork, pork raised outside, and pork with real pork taste. The niches of today are the only alternatives for tomorrow.

The Family Hog Farm

I honestly believe that nothing will be sacrificed by a return to production entirely from family farms. Note these favorable points:

- Small herds mean big litters of good pigs. Small-scale producers have the time to see that everything gets done right the first time and to know and manage each animal in the herd as an individual.
- They produce hogs as a part of a diversified plan of production. The big corporations offset startup costs and downturns in the price cycle with income from other sources; small farmers can do the same with income from other enterprises on the farm.

THE PERSONAL TOUCH

Studies have shown that the way a person handles hogs is important. Pigs that receive pleasant and empathetic handling, for instance, are easier to manage, grow faster, and perform better than those that receive unpleasant handling. It's the kind of care that the small-farm operator can provide.

- Small diversified farms have a level of environmental integrity that the big, specialized outfits can never match. The wastes from 10 sows go back to the fields with a manure spreader; but the wastes from 1,000 sows go into a lagoon that may develop problems with the next big rain.

- Small size in no way precludes quality. Pick up a 60-year-old ag text, and you will find page after page of references to "ton litters," 10-pig litter averages, and hogs hitting a 220-pound market weight at a bit older than 5 months of age — and all of this performance was fueled with open-pollinated corn, skim milk, and tankage. If small farmers did this once, they can do it again.
- Small farmers are discerning users of technology and not slaves to the methodology of the moment. They don't have to have the latest bit of techno-puffery to help them cope with the health, ventilation, or waste problem of the month.
- They can and do work with quality inputs. Time and again, breed-building boars and sows have come from small farms, and the family farmers are now the ones guarding the gene pool that the seedstock companies draw from to create their syntho-composites.

AN AMISH EXAMPLE

The ability of the small-scale swine producer to survive is perhaps best epitomized by the Amish example. West of the Mississippi, Amish and Mennonite farmers have embraced pork production nearly as often as they have dairying in the East. And they do it with a clear-eyed vision of what they are about.

Over the years, as a field representative for the Hubbard Milling Company, I worked with Amish farmers, and supplied them with purebred breeding stock. The Amish farmers I've come to know operate at levels of from 10 to 60 sows and produce feeder pigs, butcher hogs, and crossbred breeding animals. They were among the first producers in our area to regularly use F_1 breeding animals and follow crossbreeding rotations.

These farmers are very receptive to new, practical ideas that will help them obtain "optimum production." They read the hog magazines; study type trends; give new products a fair trial; invest in quality breeding stock; use simple, easily maintained housing and facilities; and use pork production to best and fully utilize available family labor. Optimum production means seeking a fair and reasonable return on investment rather than attempting to totally maximize production in the faint hope that returns will eventually surpass costs. In other words, optimum means aiming for a fair and livable return, rather than attempting to extract every cent of return possible by any means available.

I've seen $500 boars go to Amish farms with but 10-sow herds, have known Amish producers to contact breeders 100 miles from home and order boars sight unseen to get genetics they needed to move ahead (they did their genetic homework first) and seen delivered to them some of the newest in feedstuffs and animal-health products. Within the use of just a couple of hundred pounds of a new feedstuff or a test on one set of pigs, these farmers can discern if a product or method will work for them.

Defining Optimum Production

To illustrate the optimum-production concept, let's take apart the old industry chestnut that every sow has to wean at least six pigs per litter just for you to cover all costs and break even. It's still not a bad rule, but in an expensive facility with a totally controlled environment that figure is now more like 7½ pigs per litter needed to break even.

Interestingly, there are some folks out here weaning six pigs per litter and making money at it. They're farrowing in the brush, have a couple of $25 electric-fence chargers bought at auction, are farrowing in 20-year-old huts, follow a more seasonal farrowing pattern, use a well-thought-out plan of crossbreeding, and market their production rather than just trying to get something sold whenever a note payment comes due.

The swine industry is now wrestling with the question of just how many pigs a sow can produce in a year. There are producers in Europe bumping 26 pigs per sow per year, and some believe that 30 or even more is doable. Very early weaning, multistage nurseries, and sows that must be supported with red-hot diets and rigidly controlled environments make this doable, yes, but at a very great cost.

There are sow lines now that can no longer be penned outside during the changing seasons, nursery buildings run at a steady 90°F (32°C) year-round — no matter how high the cost of fuel gets or how great the environmental impact of these corporate systems is on the planet — and pigs being weaned at 10 pounds rather than the usual weaning weight of 25 pounds at 35 days or 40 pounds at 56 days. How many producers are there who can realistically afford to be this "good"?

In nearly every operation there is room for improvement, but there is also that point at which further increases in production will not carry the greater costs to produce. The time and resources needed to wring every dollar possible from a venture are simply not always justifiable on economic grounds or sustainable environmentally. And let's be realistic here — life is just too short to spend the whole of it down in the hog lots rooting out every last dollar.

Archaeological data keeps pushing back the date when hogs were first domesticated. Certainly, they have moved across this earth in the company of humans for a long, long time. Think of the suckling pig laid out on the tables of kings and the cured ham swung over the saddle of a westward wanderer: pigs and pork and people have formed quite an alliance.

Ensuring Your Future

The future for hog producers is just as great today as it has always been. There will be a place in it for pork producers of all sizes — if they will accept but a few simple tenets of the successful swine herd:

- You have to like 'em even when they don't smell like money.
- Start small, but start well.
- Grow as you learn and only to a level where you are comfortable.
- Maintain quality animals that perform well and are a pleasure to own.
- Produce them with a very specific plan of action, from breeding pen to market.

12

DAY-TO-DAY LIFE WITH HOGS

A couple of years back, I was invited to speak about my hog-raising experiences at a seminar for beginning small farmers and those seeking new ventures to add to existing small-farm enterprise mixes. On the morning program, I was sandwiched between an exotic-species broker and an ostrich farmer.

Not wishing to distract interest from the speaker following me, I excused myself following a short question-and-answer period. I was greatly surprised, however, at the number of folks who followed me out of the meeting room and continued their questions in a nearby trade-show aisle. Hogs seem to at once intrigue, frighten, and inspire those folks.

One man was eager to make a start with hogs, but his wife was concerned about odors and opposed his enterprising ideas with a tenacity that would do a pit bull proud. Another was having difficulty finding certain breeds close to his home farm. A third man was just bogged down in unfavorable "hawg" imagery. I steered the group to a nearby display of live breeding hogs. A few minutes of hands-on experience allayed many of their fears. No one was bitten or succumbed to noxious fumes. Breeders were found who would deliver purchased breeding animals or at least cooperate with delivery

to central points. And the hogs endured our intrusion into their space with all the dignity inherent in their species.

Honestly, I know of no substitute for hands-on experience, whatever the agricultural venture you are considering. A few head of hogs are themselves the best teachers of what they need. They will give you the title "hog farmer" and, in time, will give you the savvy and experience to merit it. That is the gist of this chapter: How to be a good hog raiser, a good small-farm-based hog man or woman.

Movin' Hogs the Easy Way

Actually, the most difficult and time-consuming task on a great many farms with hogs is getting them from point A to point B while maintaining a modicum of dignity and your religion. I have actually seen films of a farrowing house rigged up with a length of 12-inch-diameter PVC pipe through one outside wall at a downward angle, to speed up pig handling by allowing weanling pigs to be slid through the pipe and into a waiting truck or trailer. In a 180° turn from this attempt at a pig elevator, we had an elderly neighbor who would place a bushel basket over a sow's head and back her up and down chutes and into farrowing stalls with

These pastured sows are being fed at some distance from each other to ensure that each receives the correct amount of feed.

SAGE ADVICE

Working with hogs can also sometimes be a humbling experience. A group of us supposed "veteran" producers were sitting together at a purebred-hog sale watching three ring workers struggle to cut a couple of gilts out of a larger group of 250- to 300-pound gilts in the sale ring. One of us ventured to say that they needed more cutting gates and hurdles, another said more men were needed, and a third was of the opinion that more gates and men both were needed. A truly wise old gentleman sitting behind us and high up in the bleachers put us all in our place when he correctly surmised that what was really needed was to replace at least two of those halfway-useful ring workers with one good stock dog.

aid inside a larger pen, and, with the help of a second person, a tool for moving even a large sow or boar over short distances.

It can also be tipped over a hog, or the hog can be driven into it and the short panel wired shut behind. The lightweight unit can then be dragged by the narrow front end by one person. A second can walk along behind, urging the hog forward and quickly pushing down on a side or end if the animal balks or attempts to root beneath it. Quite often, the gentle bumping from behind with the short panel is all that's needed to keep the hog moving to a new pen or pasture.

As noted earlier, the best way to keep a hog moving is to block the way behind it. As much as possible, allow the hog to move along at its own pace and in the company of other animals to help it remain confident and at ease. It is often far easier to drive two or three hogs back off a trailer than to drive just one hog onto it.

comparative ease. Keep 'em in the dark and keep 'em moving was his philosophy.

When I pull shoats from weaning quarters to move them 300 or 400 yards to growing pens, I put them into a plastic trash can on wheels. Two or three young shoats will travel quite comfortably in this unconventional but quite inexpensive and simple-to-use "trailer." It also will hold a whole litter of very young pigs while I give them health treatments or a few boar pigs while I sort through a litter for keepers and hold the rest for castration.

For larger hogs, I often use a "farm boy" hog trailer made from one 16-foot-long hog panel and a 5- to 6-foot-long segment cut from another. I bow the longer panel into a horseshoe shape, then securely wire the shorter panel across the open side. This simple unit can be used as a small holding pen, a sorting

A plastic trash can on wheels can be used to move two or three shoats quite comfortably for a short distance.

BEWARE OF LIGHT

When you're moving hogs, it helps to think ahead and even to think a bit like a hog. Changing light patterns, intensities, shadows, and contrasting colors in their line of sight can be most disturbing to animals being driven from one place to another. Especially difficult can be the task of driving a hog from a darker barn environment into bright sunlight and vice versa.

In areas where hogs will be regularly worked or along frequently used lanes of travel, many producers line the fencing or enclosure with plywood or sheet tin to completely block out shifting light patterns and other outside distractions. Hogs will also move up solid-floored loading chutes more easily than those with slatted floors.

Hogs are wary of shifting light patterns and contrasts in flooring and other footing surfaces when being moved and may balk or turn back from them.

Managing Information

Perhaps even more challenging than getting hogs moved from pen to pen with a fair semblance of order is the task of keeping current on the hog scene.

The information age has come to the countryside with a vengeance; the material available often seems to flow over small farms in great waves. At last count we receive five monthly magazines that deal solely with hogs, three others with regular inserts that target hogs, and at least half a dozen general-interest farm magazines with extensive hog coverage. This doesn't even begin to include the various catalogs, newsletters, Extension bulletins, and books that regularly find their way to our mailbox.

I awake to market reports on the radio, eat dinner (lunch to you folks outside the Midwest and Southeast) with market commentary in the background on radio or TV, and finish the day with the TV weatherman helping me select my work attire for the next day. Phyllis, my supportive wife, knows that I have at last grudgingly acknowledged the arrival of winter when I heed the weatherman's warnings and don what she calls my "long lingerie."

In talks with those contemplating a hog venture of some sort, I find that they are too often put off by this floodtide flow of ever-changing and sometimes seemingly contradictory information. It has reached the point where a single case of Peruvian Higgledy-Piggledy disease in Taiwan can receive multimedia coverage in the United States within a matter of hours and be in the print media in detail within a very few days. Fortunately, there is no such thing as Peruvian Higgledy-Piggledy disease, and information management must now become just one more aspect of good farm management and operation.

I've yet to find a hog that can read, and I trust that if I do find one so gifted, he will be wise enough to take everything he reads in the newspaper or on his favorite hog blog on the Internet with a grain of salt. In 30 years at the hog game, I've seen a great many trends, products, and practices crash and burn within months of having been named "can't miss" in the farm press. Some have even come around again a time or two. Two that rise quickly to mind are the "flat-muscling" and "late-maturing" trends of about 30 years ago.

"Flat-muscled" hogs looked to most folks like the hogs that were once called "cat hammed" and "meatless wonders." "Late-maturing" hogs were also largely "slow-growing" hogs. Of course, change is sometimes good. For example, the new role of the Berkshire breed in producing the highly sought-after "black pork" for the Asian trade looks like something that will benefit all sectors of the pork industry. Still, it seems type changes and new production methodology often are advocated only when certain sections within the industry need something new or pricier to sell.

Count on Other Producers

I want to see what I read about being tested in the field before I buy into it. Fortunately, swine producers as a group are very open and sharing people. Many magazines actually include the addresses and phone numbers of producers mentioned in their articles. I have called complete strangers to ask them about their firsthand experiences with a certain new product or bloodline. It is also a favor I have done for others like me, who are trying to take a sounding on the latest on the swine scene. Time your calls close to 8 P.M. on weekdays (don't forget the differences between time zones), keep calls short and limited to just a couple of questions, and be sure to include a self-addressed, stamped envelope with all letters of inquiry, or ask if they have an e-mail address that you might use.

I also know a number of longtime friends and fellow swine producers whose input and opinions I rely on greatly. When the phone rings between 8 and 10 P.M. on a weekday, it's generally one or another of our group plugging in to talk about what's happening locally, regionally, or nationally, or seeking help in working through a problem of one sort or another. We try not to overextend our welcome through this practice, but these kinds of sounding boards have proven vital for figuring out what will and won't work for our operation.

BETTER INFORMATION MANAGEMENT

Here are five steps to help you better manage information.

Step 1. Invest a reasonable amount of regularly scheduled time in information acquisition and management. I have often heard it said that the really good farmers spend an average of two hours daily reading and otherwise processing the information currently available to them. It is a figure I certainly believe to be correct. Tucked into the magazine rack next to your favorite easy chair should be a large envelope with pen or pencil, highlighter, and scissors. Into the envelope can go articles or even whole issues you wish to keep for future reference. It's easier to take clippings on first reading than to stack magazines away thinking you'll get back to them at a later date. Dad and I would even make marginal notes and pass magazines back and forth.

Step 2. Do your own field testing on a small scale. Make things prove themselves on your farm, which is like no other on earth. A great many farmers have invested tens of thousands of dollars in new technology, only to see it all fall out of favor in far less time than it took to come into vogue.

Step 3. Plug in to your fellow producers for their input and experiences.

Step 4. Go see it, and then go see it again before you buy. If a practice or product or animal catches your eye, study it in detail, and then bring it onto the farm in the smallest, simplest, and easiest-to-manage way that you can. Dad once traveled across three counties looking at different types of pig creep feeders, came back home, sorted through our scrap pile, and built one himself. We used it for several years before replacing it with one of the factory models he liked, which we were then able to buy used for just $25.

Step 5. The best money you can spend is the money that buys you hard and solid information. A $25 book can often save you thousands of dollars in costly mistakes. Publications and periodicals pertinent to your particular field of endeavor are generally deductible expenses on your Schedule F federal tax form.

Do It Right the First Time

I seem to recall an old razor blade commercial that said something to the effect that to be sharp you needed to look sharp. On the farm with hogs, "looking sharp" means doing things right the first time.

It has often amazed me how that lone, short length of 2 × 4-inch board leaned up in the corner of a farm building is soon joined by a pitchfork with a broken tine, an old 5-gallon bucket, a busted tank sprayer, and a push broom with the handle flamboyantly wrapped in silvery duct tape. We all need to be more mindful of the three Ps for good farmstead keeping: Pick it up as soon as it drops or falls, put it back as soon as you're done with it, and keep it painted or otherwise maintained.

Good tools and facilities do indeed seem to improve the quality of a farm's output. They can also greatly improve the efficiency of the operation, because there is no substitute for the right tool for the job. They create a pride of workmanship that can be almost tangible. A clean, neat farmstead also inspires confidence in the potential buyers who may visit you.

I will concede that, all too often, my own shop and buildings do harbor more than their fair share of "boars' nests," but I was taught better. First and foremost is the safety factor for both the producer and his or her porcine charges. Lose a needed hand tool or cause a tear in an expensive bag of pig starter, and you can swear now and laugh later; cause an injury to yourself, a family member, or a valuable breeding animal, however, and the consequences can be quite far-reaching. A barrow I was once heavily counting on to take to a local hog show injured its eye on baling wire carelessly left dangling in the pen after it had been used to tie off a gate. Within minutes, the animal went into shock and was dead.

Step on a nail, and you may be out of action for several painful days. Let that same mishap befall your one-and-only herd boar, and you are out of business.

There are times when my management style is as loose as a bucket of ashes, so when mistakes and accidents happen, there is no one to blame but myself — the creature in our hog pens who is supposed to have the biggest brain. Every farm has those catchall spaces where old gloves and slip-joint pliers seem to migrate of their own accord, but too often these are also the spaces where producers first go to look for things as dear and delicate as livestock health supplies.

Did you ever wonder why some swine-health-care products are packaged in brown-glass bottles and others are in clear glass? The ones in the brown bottles are light sensitive. Leaving them on that handy shelf at the barn or the dash of the pickup will certainly shorten their useful life and may even render them totally useless.

A WELL-RUN FARM

One of the best-run small farms I ever visited not only had a refrigerator in an outbuilding for the correct storage of veterinary pharmaceuticals, but on that fridge door was a list of what was inside, when it was bought, when it would expire, when it was last used, and how much of it remained. A veterinarian of my acquaintance measures the skills and stockmanship of his clients by how clean they keep the tops of their bottles of injectable livestock drugs.

There is an old saying that in the deep woods where the bears and wolves reside, it is still the mosquitoes that drive back most of the intruders. The small things can indeed be the undoing of farm operations of all sizes. Do the little things in a timely fashion, and they won't grow into the crushing tasks that may eventually overturn your entire operation.

Take Time to Scratch Their Ears

Several times each day, there is nothing more important that swine producers can do than simply go out and look at and listen to their hogs. Early in the morning and late in the day (around sunrise and sunset) are excellent times to take a really hard look at the animals in your care. At these times, give your animals your full attention rather than attempting to note things simply in passing. It is at these times of the day that soundness and respiratory problems should be most noticeable. They are also good times to note the early listlessness that so often foretells developing health problems. Those animals that are slow off their beds or that distance themselves from others in the group should receive a much closer inspection and perhaps have their rectal temperatures taken, to help form an early diagnosis of sorts.

This is also a feel-good thing for most farmers with livestock. Seeing the hogs well fed and content at the end of the day can take the edge off all those aches and pains you've been feeling. The animals are your hopes and dreams for the future given form and substance, and you have every right to feel content and rewarded as you stand viewing them.

Goals

I have never met a hog farmer who was completely caught up with chores. We all have something more to do or something more we want to incorporate to add value to our swine ventures. And we've all had days when we labored mightily just to get to a point where we felt it was safe to leave the hogs for naught but a warm meal and a few hours of badly needed sleep.

SIZE UP YOUR HOGS

You need to spend time really looking at your hogs, comparing them with that mental standard or ideal that all good producers carry around in their heads. Duly note those animals that don't measure up to that standard, and mark them for elimination from the herd. An old proverb says that nothing fattens the beast like the eye of its owner; we all have to learn to look past what we want our hogs to be and see them as they really are.

Pull out that pocket herd book or draw on your memory, and start matching up pigs to their parents. Are old Sally's pigs growing like you thought they would, or is it time to start thinking about how many pounds of pepperoni on the hoof she represents? Are those feeders that cost so much to buy muscling up the way you anticipated, or is it time to look for a new pig supplier?

One of the most important things that swine producers can do is simply go out and look at and listen to their hogs.

This wanting to make things better should be an incentive and an element of pleasant anticipation within your swine venture. Don't beat yourself up emotionally for not achieving everything here and now. The story is often told of the farmer who met the visiting county agent at his gate and told him not to even bother to come in, because he already knew how to farm better than he could afford to.

I believe in the concept of stringing together small, doable goals to create an ever-improving operation. Too many farmers work only for that one big day when, somehow, weather, markets, crops, and genetics will all come together to make them a great success in one fell swoop. Buy into that one and you will burn out or die with a broken heart.

In the early going, we worked to add one or two good, home-raised gilts every few months or to add a little better boar than the one before it at intervals of 18 months or so. I raised a lot of hogs in our old horse barn before I bought my first new hog house, and I raised purebred Durocs for nearly two years before adding a

WORK SLOW AND STEADY

Slow and steady on the farm are ways to both learn and grow. Weaning an eight-pig average should make you want to try for an eight-and-a-half-pig average next time, not to invest in a $60,000 farrowing house to hold dozens or more sows to wean eight or fewer pigs per litter. A goal should be something attainable in a reasonable amount of time — not something that places the whole future of your venture in peril should it fail to materialize.

second sow to that venture. Our first boar with a recorded pedigree was bought "used" from a neighbor who was holding back gilts and no longer needed him.

Every Pig Is Precious

The real strengths that a small-scale producer has to bring to swine production are time and attention to even the smallest of details. We do not put off until tomorrow because we know that kind of thinking will take away from profits today.

The classic example of this strength manifesting itself is in the average litter size found on the nation's small farms with sows. These are the natural homes for truly big litters, because where litters are few in number, every pig is precious, and efforts to save them are unstinting.

Cross-Fostering Practices

To increase output from the farrowing unit, many producers follow a practice called **cross-fostering**. They place nursing pigs with sows and in litters where they will have the greatest potential to thrive. This is also one more reason for farrowing females in closely timed groups.

For the sake of an example, let's take a hypothetical producer with 12 sows divided to farrow into two groups of 6 sows each. He breeds each group of sows to farrow within a very few days of each other.

He enters his farrowing house early one morning to work with the six sows there, and finds that all six had farrowed within 120 hours of each other.

Three have good litters of 8 to 10 evenly sized pigs. One is an older, heavy milking sow with just 7 large pigs but with a good track record as a pig raiser; another is a second-parity sow with 11 rather unevenly sized pigs; and the last is a gilt with 6 big pigs and 2 smaller ones. By making a few number adjustments among the latter three litters and allowing some pigs to be raised by dams other than their own, the producer can wean a more even crop of pigs, increase pig survivability, get optimum production from all sows, and breed back these six sows as a uniform group for the next farrowing.

Here are 10 tips to successful cross-fostering:

1. Don't switch pigs receiving colostrum to sows no longer producing colostrum and vice versa.

2. Try to switch only those pigs born no more than 36 hours apart.

3. Contrary to what many might think, the best pigs to move from litter to litter are not the small ones but the largest and most vigorous pigs in the litter.

4. The bigger, more vigorous pigs should go to a sow that can milk well, to maintain their current level of performance.

5. Small pigs should be grouped with a sow that has a proven track record as a pig raiser and as long a potential lactation period before her as possible.

6. Get litters as even as possible while moving as few pigs as possible.

7. Handle the little pigs very gently, to keep them from squealing or showing other signs of distress when they are placed in a new pen and with a foster dam.

8. Try to place little pigs on the new sow while she is nursing.

9. Monitor the newly restructured litters for several hours following relocation of the pigs. Any not making the transition can be returned to their dams.

10. In cross-fostering, many believe that scent can be a factor in getting a sow to accept her new family. Toward this end, a lot of little pigs have been liberally doused with those Christmas talcs and aftershaves that most of us have stored away over the years. These lend some interesting new odors to the farrowing quarters. Still, I believe that the normal tumbles and turns of a litter of little pigs soon has them all smelling the same anyway.

Don't Pack 'Em Too Tight

I pointed out earlier the need to have a place for every tool and to always return it to that place. The same holds true for the hogs themselves. I believe that on even the smallest of farms there can be room for several livestock and poultry ventures of modest scale—but not all in the same place.

Many years ago, there was a book that had as its central theme the maintaining of a great many livestock varieties in a single building or barn. The appeal of such a practice is readily evident, but how successful this single-site loading would be in the real world is quite another matter. We have hog houses that have been used—at different times—to house everything from a sow and litter to growing hogs to ewes with lambs to baby chickens, but never in any combination.

In a shared barn, little pigs just don't last very long beneath the stamping, kicking feet of cattle and horses. And small calves and large hogs don't make good penmates or neighbors.

Poultry ranging free through a barn can spoil bedding and feedstuffs, damage equipment with their droppings, and unsettle some hogs with their activities. Over the years, I have heard many reports of sows that developed a strong taste for chicken tartare. Chickens 1 year and older can also harbor the avian tuberculosis germ, which may be transmitted to the hogs.

Everything in one barn sure sounds like a good idea, but it was a practice not even pursued in pioneer times, when livestock housing was at an even greater premium than it is now. Those early farmers knew that hogs, cattle, and other livestock species would fare better on free range than when crowded into a single facility.

Buying at Auction

Throughout this book, I have made reference to buying quality used hog equipment. One of the best places to find such equipment is at a farm auction. It is a fast-paced atmosphere in which to attempt to make purchases, however.

Over the years, I have assisted my professional-auctioneer in-laws in conducting a number of farm sales they booked. Through observation and from working the ring at a good many farm sales, I have picked up a few practices that help ease the chore of buying at auction:

- Carefully study the sale bills, to determine if the offering contains items that will fit your needs.
- Try to visit the sale site a day or two before the actual auction. You can then take longer to examine in detail the items in which you are interested. The owner will also have more time to answer your questions about how the items were used and maintained. When you're buying used livestock equipment and housing, it's especially important to determine if the equipment has been recently exposed to any disease-causing organisms.
- Arrive early on auction day, and dress appropriately. Don't wear the clothing and footwear you usually use for chores. Hog owners and their vehicles from all over may be in attendance, so the potential for contact with any number of potentially harmful disease organisms is great — many of these can be transported on boots, clothing, and the undercarriages of vehicles. And dress for the weather. Often sales have to go on whether the snow flies or the rain falls, and those dressed to stay and be comfortable to the very end may reap some real bargains.
- Don't start bidding without a fixed top bid already in mind. Don't go above that bid no matter how much the auctioneer and his crew wheedle and cajole. An old rule of thumb that serves many well is to spend no more than half of their new price for used farm goods.
- Don't hesitate to bid first if the offering is something you really want. People are reluctant to start bidding for fear they'll look too eager, but I have seen far too many items fall to the first bidder in my effort to hold back on something I want. Bidding first may be the only chance you have to put some kind of definition on the bidding range.

- In many instances, it's a good idea to offer a bid somewhat below the one the auctioneer is seeking. If the auctioneer has a bid of $10 in hand on a nose snare, is asking for $15, and the bidding has slowed down, it may be time to offer a bid of $12.50 or even $11. A time-honored bidding gesture is the slashing of one index finger across the other to indicate a bid increase of just half of the one the auctioneer is requesting.

- Step back when you've reached your bidding limit. This will shake the auctioneer off and let it be known that you're done. Don't feel self-conscious about standing by your limit. Most veteran auction goers will respect you for your savvy approach and well-thought-out bidding.

Hog-Farm Fashion

Fashion might seem like a funny thing to talk about in a book about hogs, and I will grant that, rather than fashion plates, most hog farmers are likely to be fashion saucers. Still, I have already noted the importance of proper dress for attending an auction, and for all of your tasks, the right clothing and footwear can do much to add to your comfort and safety.

For example, there is far more than simple mud and muck in spring and fall hog lots. These lots are liberally laced with urine and manure, which can take a toll on even the best shoes and boots. Further, when your feet are wet, chafed, or otherwise discomforted, the rest of you is out of sorts, too. If you regularly wear leather footwear, make sure to buy work shoes that have been treated to be manure and urine resistant. There are also a number of leather-shoe treatments that will help maintain their suppleness and resistance.

IT'S AS IS!

Remember, the merchandise at an auction usually is sold "as is." In fact, the terms of sale at most farm auctions are "as is, where is." If you buy it, then go to pick it up and the bottom falls off, you'll learn to look more closely next time. Any expressed warranties are strictly between the buyer and the seller; the auctioneer is the agent for the sale and nothing more. If the seller wishes to hold a reserve bid or floor price on an item, this must be announced as the item goes on the block or it has to be sold to the highest bidder. Following the sale, you will have from 7 to 30 days to move the goods from the premises. Short days should be announced by the auctioneer as a part of the terms of sale. The auctioneer should also be able to assist you in locating truckers to haul your purchases if needed.

When the hammer falls and the auctioneer announces an item sold to you, it is your baby 100 percent. Should it be stolen, run over in the parking lot, or sat on by an elephant, it's your loss to take. Sadly, you are more and more apt to encounter theft in even the most rural areas these days, and the smaller items at auction seem to be especially vulnerable to theft in all of the confusion.

Author Kelly Klober in his current-day work clothes

There have been years — 1993 and 1994 come to mind — when it seemed that I wore knee boots every day except the Fourth of July. The slip-on styles are easy to get on and off, and they do a pretty good job of keeping your lower legs dry, along with your feet. You can spend a lot for a pair of knee boots, but due to the ever-present risk of tears and punctures while doing farm tasks, I prefer to wear less-expensive boot models. It seems to be the Klober law of rubber-boot ownership that the more expensive the boots are, the more likely I am to punch a hole in them.

When I first became involved with pure-bred hogs in the late 1960s and early 1970s, many of us affected a sort of British herdsman look. We wore green knee boots, a coverall with full sleeves topped by a khaki or duck coat or vest, and on our heads a short-billed, brown-duck Jones cap. The back of that cap could be snapped down to keep rain from going down the back of the neck.

It was warm, it was durable, and by golly, we looked like hog farmers. It was a style of dress that endured until the "gimme" advertising caps appeared on the scene and we opted for a denim-rich, flannel-heavy "urban farm boy" look. We still hung onto our rubber boots, and Dad often teased that I should quit worrying about supposedly free caps, go where the feed was the best buy, and use part of the money I saved to buy my own headwear. It was good advice.

Denim and flannel still dominate today's choices in work clothes, but producers in temperature-regulated buildings are opting more and more for short-sleeved coveralls (I can't bring myself to call them jumpsuits). It's sort of like Star Fleet designed a uniform for the manure-spreading set.

All kidding aside, good work clothing that fits well and is made of durable fabric is one of those small but very important things that ease your labors (and it is also a deductible

THE CREASE IN THE CAP

Midwestern farmers do add personal touches to all those multicolored, mesh-backed caps given away by feed and farm-supply companies. You can sometimes get a clue as to where they hail from in the Corn/Hog Belt by the way they crease the bill of their caps. The sharp crease in the center of the bill seems to be the Missouri look of the moment; across the big river in Illinois, they appear to favor two creases, on the outer edges of the bill, giving it a squared-up appearance.

Thus far, hog farmers have avoided those dinner-plate-size belt buckles and reptile-skin boots, but I would like to share a short joke that every winter seems to make the rounds of all those supper meetings the feed companies use to introduce their new products. Do you know why hog farmers don't wear athletic shoes? Because feed companies don't give them away.

expense). Very loose-fitting clothing can be a safety risk, because it can be more easily grabbed by spinning shafts and moving gears and can pull you into them. The traction and grip provided by quality footwear also add greatly to your safety.

Hog People

There is no finer group of people on the land than this nation's family farmers with sow herds or related ventures. Invariably, I have found them to be considerate and sharing people, willing to give freely of their time and experiences.

They know what it is to live high on the hog and low on her hocks — yet I have never met a single producer who was hardened or had a real root-hog-or-die attitude toward life. When you enter their ranks, you will find yourself a part of a real fellowship based on shared experiences and a commonality of goals.

You will have to earn the mud on your boots and the dings in your pickup, but you are part of an industry that was spawned with Columbus's second voyage to the New World. You may cuss 'em up the loading chute one day and cry over them the next when pigs in a promising litter are lost despite your best efforts. But then, that's farming.

They don't sing songs about hog farmers the way they do about cowboys, but those of us with mud on our jeans and hogs on our farms can take a great deal of satisfaction in knowing that in this protein-starved world, we're producing one of the leanest, most healthful, and most nutritious of all foods.

I like hogs. I've also made a lot of payments with their help and met a lot of good folks through them. I still see in them a promising future for small farmers willing to tend them in a thoughtful and caring way.

People with the time and talent to be good hog raisers also seem to have the best of human virtues, and indeed, they make the best of neighbors.

Many in economic and marketing fields believe that we may actually see $1 per pound pork futures contracts traded before too long. It is certainly nothing I ever expected to see, but then I didn't expect 8-cent hogs back there in the 1990s either.

We have yet to see the former figure, and it is always possible that the latter may occur again. It is very much a changed time for swine production and swine producers. From hard times and what some have termed fragmentation in the industry have come new opportunities and responsibilities.

The small producer remains, and what a victory that is. Our role is very different from that of even a quarter of a century ago, but it is very much in keeping with the historical role of the small family farm and farmer and American pork production and swine raising.

By becoming more focused on the care and breeding of the hogs, small producers are finding a way out of the morass of agribusiness and pork as a mere commodity. After years of drifting away, there are young people and families coming back to raising hogs. It is happening on a modest scale but with the solid backing of a dedicated consuming population that approves of what we are doing. Consumers are seeking input now, and rewarding our production with premium prices.

I feel better about this kind of production than I have for some time and see a very bright future for this truly artisanal type and level of production agriculture. Ultimately, it will be what we, the individual producers, make of it. We have an opening unique and apart from any seen in my lifetime.

With an heirloom breed, a special product, a distinct approach to production, or a combination of all three, a better and more becoming life can be created for both the producers and the hogs they now raise.

HOG ANATOMY

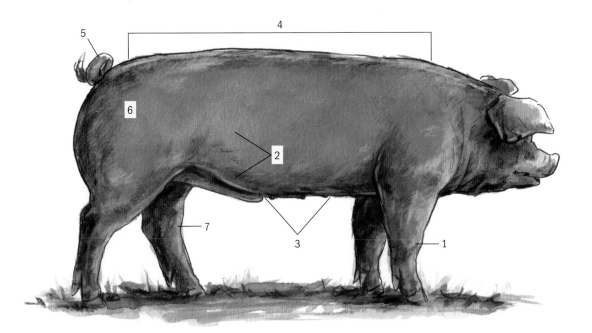

This is the idealized market barrow. Notice the well-formed front legs (1), the depth of the side (2), the three nipples (3) ahead of the penile sheath, the long and level topline (4), the well-positioned tailhead (5), the well-defined ham (6), and the good bone diameter as reflected in leg size (7).

REPRODUCTIVE INFORMATION

Here is some basic reproductive information, to help you determine times of estrus, gestation, and farrowing.

Reproductive Facts

- The duration of estrus is one to five days, and the average is two to three days.
- The regularity of estrus is 18 to 24 days; the average is 21 days.
- The first occurrence of estrus after farrowing will take between one and eight weeks but can then be expected to occur every 17 to 24 days until the sow is bred. Two litters per sow per year is a reasonable goal, and some producers are achieving five to six litters in a 24-month period, with weaning of the pigs at 5 weeks of age or less.
- Gestation takes between 110 and 116 days; the average is 113 to 114 days.
- Puberty for gilts can occur between 120 and 235 days of age; the average is 200 days of age. There are some differences between breeds as to the onset of puberty in gilts.

GESTATION TABLE

This information will help you determine when farrowing will occur.

Breeding Date	Farrowing Date	Breeding Date	Farrowing Date
Jan 1	Apr 25	Jul 1	Oct 23
Jan 15	May 9	Jul 15	Nov 6
Jan 30	May 24	Jul 30	Nov 21
Feb 1	May 26	Aug 1	Nov 23
Feb 15	Jun 9	Aug 15	Dec 7
Feb 28	Jun 22	Aug 30	Dec 22
Mar 1	Jun 23	Sep 1	Dec 24
Mar 15	Jul 7	Sep 15	Jan 7
Mar 30	Jul 22	Sep 30	Jan 22
Apr 1	Jul 24	Oct 1	Jan 23
Apr 15	Aug 7	Oct 15	Feb 6
Apr 30	Aug 22	Oct 30	Feb 21
May 1	Aug 23	Nov 1	Feb 23
May 15	Sep 6	Nov 15	Mar 9
May 30	Sep 21	Nov 30	Mar 14
Jun 1	Sep 23	Dec 1	Mar 25
Jun 15	Oct 7	Dec 15	Apr 8
Jun 30	Oct 22	Dec 30	Apr 23

RECORD-KEEPING TEMPLATES

Pedigree Form

You can copy this blank pedigree and use it for your hogs, or go to www.storey.com/SGTR-Pigs and download a PDF. Record the ear notch of your pigs to avoid inbreeding by positively linking the pig to its sire and dam.

Breeder _____

Address _____

Telephone _____

Sire _____
Ear Notch _____
Color _____
Wt. _____
Winnings _____

Sire _____
Ear Notch _____
Color _____
Wt. _____
Winnings _____

Dam _____
Ear Notch _____
Color _____
Wt. _____
Winnings _____

Breed _____
Born _____ Sex _____
Name _____
Ear Notch _____
Color _____
Wt. _____
Winnings _____

Sire _____
Ear Notch _____
Color _____
Wt. _____
Winnings _____

Dam _____
Ear Notch _____
Color _____
Wt. _____
Winnings _____

Dam _____
Ear Notch _____
Color _____
Wt. _____
Winnings _____

Sire _____
Ear Notch _____
Color _____
Wt. _____
Winnings _____

Dam _____
Ear Notch _____
Color _____
Wt. _____
Winnings _____

Sire _____
Ear Notch _____
Color _____
Wt. _____
Winnings _____

Dam _____
Ear Notch _____
Color _____
Wt. _____
Winnings _____

Sire _____
Ear Notch _____
Color _____
Wt. _____
Winnings _____

Dam _____
Ear Notch _____
Color _____
Wt. _____
Winnings _____

Date _____

Sold to _____

Address _____

I certify that this pedigree is correct to the best of my knowledge and belief.

_____

RECORD-KEEPING TEMPLATES

Monthly Management Chart

Post in a location that will be convenient to record the facts about your hog raising. Monthly sheets can be totaled to determine an "annual report" of your hog production. Go to www.storey .com/SGTR-Pigs to download the PDFs.

EXPENSES

| Date | Animals Purchased | | Pounds of Feed Used | Cost of Feed | Other Costs: Supplies, etc. | |
	No.	Cost			Item	Cost
Total						

INCOME

(List income from sale of equipment, breeding fees, etc.)

Date	Item	Amount Received
	Total	

BREEDING RECORD

| Date Bred | Name or Number | | Date Due to Farrow | Date Followed | Number of Live Young Born |
	Boar	Sow			

Total

SHOW PARTICIPATION

Date Won	Name and Place of Show	Placing or Award Received	Entry Fees	Value of Premiums

Total

SAMPLE CALENDAR YEAR FOR THE DIVERSIFIED SMALL PRODUCER

The fully diversified producer must sit down with pencil in hand and carefully lay out the farming year ahead (sometimes more than one year). Ventures can indeed grow out of each other but must never compete directly for the same space, producer time, or other farm resources.

You can know the dates and plan nearly four months ahead of when sows are due to farrow. Do not, then, breed them to farrow when you should be planting corn or selling tomatoes two to three days a week at area farmers' markets. Nor do you want to have 200 broiler chicks arrive the same week you are weaning calves or pigs. It is also not a good time to have lambs arriving when strawberries are ripening.

The following is a rough schedule of what might be expected to unfold for a diversified, mixed-stock producer in the Midwest during the course of a year. It does not note daily tasks such as feeding and health care, but rather points out the busy seasons for the various species and how they might interact with and around various swine ventures.

Ventures will vary from farm to farm and from region to region, but this calendar offers options for selecting and scheduling ventures in a noncompetitive way.

January

This is deep winter here and generally a time of planning, tax work, and maintenance and cleaning around the farmyard. It is also time for a bit of rest and to catch up on some reading. It's catalog season, and the wish books should inspire some thinking about the year ahead.

Show pig and specialty pork raisers may have some early litters arriving. Clean out the brooder house or units, mend lambing jugs, adjust rations for the weather, and do other shop work. A few will have early lambs and kids arriving for the Easter trade. Early in the year, births will require lots of time and care and special quarters and must be planned for accordingly.

February

Do more farrowing early in the month if producing show pigs or breeding stock. Start kidding goats, lambs, or pullets for the laying flock later in the month.

Begin readying equipment for spring fieldwork, frost-seed pastures (seed frozen crops with clover or other pasture species), and mine those farm records for tax time and future planning. This is the beginning of the busy season for greenhouse and hoophouse growers, so plan carefully to keep free the blocks of time needed for these ventures.

March

Spring is here in fits and starts, but this can be a time of deep mud and cold rain, too. Seasonal swine producers will have litters arriving after mid-month. Goat and sheep farmers should consider the timing of Islamic holidays.

Start broiler chicks a couple of weeks ahead of farrowings, or schedule kiddings or lambings ahead of farrowings. Renew contacts with buyers of your direct-marketed items from sausage to tomatoes. Time to wean and market those early show pigs.

April

It's spring proper and some calves will begin arriving for those following a more seasonal system of production. There will be more pigs to wean and possibly more chicks to start.

Fieldwork (tillage and some planting) can begin in many places, and livestock tasks should be kept to a minimum after the middle of the month. Use rainy days for cleaning and maintenance chores and to take the time to walk the herds and flocks twice a day to monitor health and condition. Dad used to say that if you couldn't walk the whole of your farm before breakfast, you were farming too big.

May

Once those oak leaves are as big as squirrel ears, fieldwork begins in earnest. Planting, and later haying, will take up most of the month. Farmers' market season is beginning, bulls are being turned back to breeding pastures, turkey poults and waterfowl can be started for their seasonal markets in the fall, and it is time to kindle some litters for youngsters needing meat rabbits to show at the summer fairs.

Seasonal calving is concluding, sows are being bred or are in early gestation, ewes and late lambs are on pasture, and the first broiler chicks may be harvested late in the month. Do take the time to monitor breeding pastures and lots, and make sure that young males and females get off to a good start and are maintaining condition. There are never enough days in May, and it is the month when even the best of plans may jump the rails and crop and market garden work must have top priority.

June

Hopefully, hopefully, the crop is in, but cultivation work must now begin. Work picks up in the garden, there may be hay to mow and bale, and many try to build some fence before it gets too warm. Haymaking and late plantings must receive precedence, and some wheat harvest can even begin late in the month.

There may be broilers to start, piggy sows will need monitoring, and it is time to begin preparing for the upcoming hot weather. Farmers' markets really pick up, and more time will be needed for direct marketing, although some pork sales may decline in warm weather. It is barbecue season, however, and pork steaks and brat-type products will be in demand. (Pork steak season in Missouri runs roughly from January 1 to December 31.)

July

Warm weather is here, and statistically, our warmest days run from mid-July to mid-August, with the hottest days in late July. Some litters will be arriving for those on a two-litter system producing show pigs.

The market garden remains a busy place, and broiler sales should be increasing. Most hoof stock is out to pasture, and the laying flock may need extra care to cope with rising temperatures to remain productive.

August

More litters are arriving, and breeding begins for early kids and lambs. Another set of broilers may be started. Heat will slow growth and reproductive performance.

The produce flow really begins here, and it is also time to begin prep work for the fall harvest. Still, this can be the month to take a bit of a breather, and some family days are always time well spent.

September

Seasonal swine producers begin farrowing again, and sheep and goat breeding continues. Early-started pullets are approaching laying age, and it is time to start one last set of broilers or roasters for the holiday markets. Breeding for early spring show pig litters begins, too.

Harvest work can begin, and a last cutting of hay may be possible in some areas. Going into the fall, proper harvest work must be the primary focus and some stock chores put off to rainy days. This is another busy month in the market garden, too.

October

Harvest and the last of the seasonal farrowing occurs. If time demands are very great, breed only older, more experienced sows to farrow in these seasons. Free time (free time?!) can be given to winter preparations.

Begin or at least get set for calf weaning and marketing. Fine-tune the laying flock and housing for winter production. Get ready to process and market turkeys and waterfowl for holiday shopping.

November

This is the traditional month for marketing feeder calves and starting calves or shoats in the feedyard. Set cows or sows to gleaning the fields following harvest. Dress and market heavy fowl for the holiday market.

Do winter prep, lay in health supplies, plan end-of-the-year measures for tax management (special purchases, buying added feed), and take time to give thanks.

December

Take stock of the year, study farm records, cull failing ventures, log out a plan of action for the coming year, and take time to investigate potential new ventures.

December sometimes allows for some added barn cleaning and manure spreading, and harvesting the last of the heavy fowl. Meat sales should strengthen going into cold weather.

Granted, nature's timetable is far from railroad-watch exacting, but it does allow for some rather fine-tuned planning. Much can be achieved in a smooth and timely fashion if the farm and the producer are not overly taxed. Think all things through completely, as pursuits begun today may have far greater and more pressing needs 6, 9, or 12 months from now. If you've got little pigs and baby chicks sharing the same heat lamp or juvenile turkeys and feeder pigs running ahead of the rotary hoe, you have a bit of a hole in your management plan.

I have emphasized throughout this book the importance of keeping current on swine trends and staying in touch with others in the industry. Here are some resources to help you toward that end.

Periodicals

Acres USA
800-355-5313
www.acresusa.com

Center for Rural Affairs
402-687-2100
www.cfra.org
A newsletter surveying events affecting rural America

Countryside & Small Stock Journal
970-392-4419
www.countrysidemag.com

Farming Magazine
330-674-1892
www.farmingmagazine.net

Missouri *Farmer Today*
800-475-6655
www.missourifarmertoday.com

National Hog Farmer
800-441-1410
www.nationalhogfarmer.com

The Progressive Farmer
800-292-2340
www.progressivefarmer.com

Purple Circle
806-499-3749
www.purplecircle.com

Seedstock EDGE

National Swine Registry
765-463-3594
www.nationalswine.com

Suppliers

Nasco
Fort Atkinson, Wisconsin
800-558-9595
www.enasco.com/farmandranch

WXICOF
Wentzville, Missouri
636-828-5100
www.wxicof.com

Internet Sites

For those of you who are computer savvy, the web will provide a vast array of information. Use one of your search engines to search for "swine" and you'll be amazed at all the resources you can access. Here are some of the websites that are available.

Animal and Plant Health Inspection Service

US Department of Agriculture
www.aphis.usda.gov

Homestead.org
www.homestead.org

This website has everything from information for producers to information on the nutritional value of pork, as well as recipes and special sections for kids.

United States Department of Agriculture (USDA)
www.usda.gov

Breed Associations and Registries

American Berkshire Association
West Lafayette, Indiana
765-497-3618
www.americanberkshire.com

American Guinea Hog Association
Naples, New York
www.guineahogs.org

Certified Pedigree Swine
Peoria, Illinois
309-691-0151
www.cpsswine.com

A unified organization of the Chester White Swine Record, National Hereford Hog Record Association, National Spotted Swine Record, and Poland China Record Association

National Swine Registry
West Lafayette, Indiana
765-463-3594
www.nationalswine.com

A consolidation of the American Yorkshire Club, the Hampshire Swine Registry, the American Landrace Association, and the United Duroc Swine Registry

The Red Wattle Hog Association
Horse Cave, Kentucky
www.redwattle.com

Miscellaneous Web Addresses

Here are the names and websites of organizations that may come in handy and be another good source of information for you. Registries or clubs for the leading swine breeds also appear below; others can be located through the National Swine Registry.

American Association of Swine Veterinarians

515-465-5255

www.aasp.org

Appropriate Technology Transfer for Rural Areas

National Sustainable Agriculture Information Service

800-346-9140

www.attra.org

The Livestock Conservancy

919-542-5704

http://livestockconservancy.org

Cooperative Extension Service

For more information about hogs and programs in your area, contact the Cooperative Extension Service in your state. This program is affiliated with each of the nation's land-grant universities and the U.S. Department of Agriculture in Washington, D.C., and can provide information on a wide range of topics.

To find your nearest Extension office, contact:

Cooperative State Research, Education, and Extension Service

United States Department of Agriculture

Washington, D.C.

202-720-2791

www.usda.gov/topics/rural/cooperative-research-and-extension-services

additives. Additions to a ration that is blended to be nutritionally complete. They can include vitamins, minerals, probiotics, medicants, and flavorings. Their main purpose is to promote growth or fully balance a ration; second, to prevent health problems from occurring; and third, to maintain economically acceptable performance in the presence of disease. Additives are generally mixed into the ration at a very low rate — often between 1 and 5 pounds per ton of a complete feed.

amino acids. These are the nitrogenous compounds that are so often referred to as the building blocks of protein. In an effort to better fine-tune rations, many feed companies now seek amino acid balances in their rations rather than formulating feeds on the older crude-protein basis.

anthelmintic. An agent that gets rid of intestinal worms.

antibodies. Bodily units of defense that rally against infection.

atrophic rhinitis. A respiratory disease that can facially disfigure the animal, make it more prone to pneumonia, and reduce performance.

baby pig. This term refers to the young pig in its first 14 to 21 days of life, when it is still dependent on a liquid diet, has yet to develop any natural resistance, and cannot generate sufficient body heat for survival.

barrow. A castrated male. He is the meat animal that is the basis of the pork industry, although gilts do produce leaner carcasses.

bin. A round or square container made of wood or metal and used to store feed, grain, and feed components. It should provide protection against moisture, birds, and rodents.

bloom. A visual indication of good health, growth rate, and feed consumption. Haircoat will be bright and smooth; the body will have a plump, full appearance; and the pigs will be bright-eyed and alert. A pig with a head appearing coarse or too large for its body is most likely stunted for its age.

blown apart. This term indicates good internal body dimension throughout the animal's entire length, which is associated with hardiness and the ability to grow well.

boar. An intact male used for breeding or held for that intent. The term covers intact males of all ages.

boar taint. An unpleasant taste and odor affecting the meat of mature, intact males.

body capacity. This term refers to the internal dimensions of the hog. A deep side; a deep, wide chest; and length of side are all deemed important to good health, growth, and feed utilization.

bone. This term is used to describe skeletal dimension as exhibited in leg diameter, size of forearm and jaw, and width in the head (which is believed to translate into width throughout the entire body). Bone is important for durability and is the frame on which muscle — red meat — hangs.

botulism. The disease complex that emerges from meat and feedstuff contaminated with *E. coli*; food-borne illness.

bred gilt. A female bred for her first litter.

bulk. This term is often used to describe feed sold by the ton or hundredweight rather than in the sack. A ton of feed to be delivered to a bin or feeder is said to be in the bulk. Bulk may also be used to mean fiber added to a ration for improved digestibility purposes.

butcher. This refers to a hog being readied for or sold on the slaughter market. A butcher hog is the same as a market hog; it weighs from 220 to 260 pounds and usually is 5 to 7 months of age.

carcass. The dressed body of a hog. At this time, the hog can be most accurately evaluated for health and meat type because this is when the ratio of fat to lean and the percentage of lean cuts (ham, loin, shoulder, and butt — the cuts of greatest value) that the animal produced can be determined.

charger. The device that controls the electric current in the fence line.

chilling. The cooling — with adverse effects — of a pig or hog by rain, wind, damp bedding, drafts, or mud.

chitterlings. The small intestines of a pig.

clean. In one instance, this term refers to a hog with visual indications of little fat cover: A hog with a trim jowl, for example, might be described as clean-headed. Clean also refers to ground or facilities that have had no contact with hogs for an extended period; six months is a minimum time for ground to be considered clean, because most disease and parasite cycles should have been disrupted within that time frame. Clean also refers to the act of the expulsion of the placenta by a sow that has just given birth.

cob rollers. Thick-necked animals whose bellies rubbed the ground, setting corn cobs to rolling.

colostrum. The first milk after farrowing, through which the sow is able to impart some of her natural immunities to the nursing pigs. It is produced for the first 72 hours following farrowing; without receiving it, the pig has almost no chance of survival. Cow colostrum has been used as a substitute, but results are still far from the same.

commodity. On our farm, we use the word "commodity" to describe industrialized, factory-farm pork.

Community Supported Agriculture (CSA). A farm or small group of farmers that directly supplies consumers with shares of the farm's harvest on a regular basis. On some farms, shareholders assist with the produce planting, harvesting, or preparations for pickup.

confinement. Containment of breeding or growing hogs in a constrained area. Now generally done inside a building with controlled temperatures.

controlled-environment nurseries. Sometimes called hot nurseries, these are early-age confinement nurseries with pigs weaned at 10 pounds and sometimes as early as 10 days. Pigs raised in this kind of environment don't transfer well to an outdoor operation as they are vulnerable to temperature variability.

coon-footed. Flat-footed, resembling the gait of a raccoon.

cover. The fat that the hog lays on beneath the skin as it approaches market weight. Some fat cover is essential for the market, but it is the most expensive gain to put on a hog, and excessive cover will be discounted.

creep feed. The early feeds, high in sugar and milk proteins, offered to a pig while it is still nursing. These feeds are very complex in their formulation and are most often processed into very small pellets or crumble, to make them easier for the pigs to consume. Walk-in feeders that protect such feed from the sows are called creep feeders.

crossbreeding. The mating of hogs of different breeds to capitalize on the strengths of various breeds.

cross-fostering. Relocating young pigs to another female to be nursed and raised by her.

crude protein. This term collectively refers to all the nitrogenous compounds in feedstuffs.

culling. The removal of animals from a herd, usually due to poor performance or illness.

dam. The female parent.

daylight. Length of leg, regarding how much daylight is seen under an animal as it stands or walks.

days. This is a shortening of the term "days to market" and refers to the time from the day of birth that it takes an animal to reach 220, 230, or 240 pounds. This can be determined anytime the animal begins to approach 200 pounds, because rate of gain can then be determined and future gains projected. When this is done, the result is referred to as "adjusted days." "Actual days" refers to what the animal weighed on an exact day in its life (most often between the 150th and 180th day of life).

direct marketing. Selling directly from the farm to the end consumer or a processor of specialty products.

drove. A herd or group of pigs.

dry cure. The curing process that uses dry seasonings in a rub form.

drylot. A large lot used to maintain hogs. Drylots are dirt surfaced but, due to foot traffic and numbers present, are free from vegetative growth. A mature hog should have a minimum of 150 square feet of space in a drylot — the more the better.

efficiency of scale. The advantage large producers enjoy when buying inputs, handling overhead costs, and marketing services in great volumes.

electrolytes. Mineral salts that increase the body's uptake of water and other liquids. They are used to counter problems with dehydration due to fever, scours, or stress.

embryo. A pig in its earliest stages of growth within the sow's uterus.

estrus. The period when the female will accept the male and conceive.

extruded soybean meal. Soybean meal created by passing soybeans through an extrusion screen. Lower in protein and higher in fat than regular soybean oil meal.

eyeball. To visually appraise an animal.

F_1. The first-generation offspring of a cross between animals from two separate pure breeds. These are the animals with the greatest boost of hybrid vigor.

farrowing. The act of giving birth.

farrow-to-finish. Taking pigs from birth to slaughter.

fed out/feed out. Fed to market weight and size.

feed conversion. Amount of feed needed to produce a pound of gain.

feed efficiency. The amount of feed the animal consumes to gain 1 pound. This figure is most often obtained as a pen average.

feeder pig. A young pig — most often between 40 and 70 pounds — produced by one farmer and sold to another for feeding out to market weight.

FFA. Once the acronym for Future Farmers of America; now the letters alone denote an organization of high school students of vocational agriculture.

finish. A term used to describe good fat cover and bloom on a hog ready for market.

finishing. Feeding out a hog for slaughter, whether it's for your family table or the slaughter market.

fitting. The process of readying an animal for exhibition.

flushing. Providing all the feed females desire prior to breeding. Done to improve fertilization and litter size.

footing. A surface that will allow the hogs to walk and mate safely and securely.

frame. The height and width of the skeletal system. A good frame gives the hog a large, massive appearance of width, substance, and durability.

free-choice feeding. Offering the animal all it wants to eat in a self-feeder.

freemartin. A congenitally defective female that will breed repeatedly but not conceive.

full feeding. Allowing the hog to consume all the feed it desires daily. This will generally be 2 to 3 percent of the animal's liveweight.

futures. A marketing arrangement whereby the producer sells a set number of hogs for future delivery on a specific date to obtain a guaranteed price for them. This can enable a producer to lock in a profit, but the numbers involved shut out most small producers. A great many speculators are involved in futures markets. I'm like Will Rogers in thinking that in order to sell a hog you should've raised that hog.

genetic potential. The inward genetic makeup of an animal, which allows it to improve the performance of its offspring above and beyond the current herd average.

genetically modified organism (GMO). Plant stocks created by genetic modification to control pests or otherwise change the plant's nature.

genotype. The genetic makeup of an animal. It's what the animal can do for you in areas such as growth rate and carcass type.

gestation. The period of pregnancy, between 110 and 116 days, with the average being 113 to 114. A lot of old-timers express it as "3 months, 3 weeks, and 3 days."

gilt. A female that has yet to bear young. The term "second-litter gilt" is used to describe a young sow carrying her second litter, but this is little more than a marketing gimmick.

grading. The sorting of feeder pigs or butcher hogs by their quality. This is done by government employees independent of the marketing concerns. Grading is an effort to reward producers with a higher premium for hogs and pigs of higher quality and greater worth.

grind and mix. A process whereby whole grains are ground and mixed with supplements to form a complete ration with bite-after-bite consistency.

gripping tongs. A hog-restraining device that applies pressure to the back of the neck.

growout period. The time it takes for a pig to grow from weaning to harvest.

growthy. Showing frame and dimension indicative of good growth.

hand breeding. Direct producer supervision of matings without allowing the boar to run with the sow herd. The desired sow and boar are placed together and returned to separate pens after mating.

hanging weight. The weight of a hog carcass.

he-boar. An especially masculine boar.

hernia. Also called a rupture; a portion of an organ or body structure that has broken through the wall that normally contains it.

heterosis. Hybrid vigor, or the increased hardiness and capacity for growth exhibited by the offspring of two different purebred strains.

hog. A swine weighing more than 120 pounds.

hog forty. A 40 percent crude-protein supplement used widely to formulate swine rations.

hogging down. Turning hogs into a field to harvest standing grain or harvest wastes.

hot nursery. A place where very young pigs undergo weaning. Pigs are held in small pens, and the building is kept at 70°F (21°C) or warmer.

hot ration. Very high protein rations.

hot wire. Another term for electric fencing — and well deserved, as anyone who has ever brushed against one can attest.

hovers. Three-sided boxlike enclosures that fit in and over a portion of the pig bunk. They are accessible only to the pigs through small "pop" holes and help keep little pigs warm. They fit over a heating pad, or an electric heat lamp is placed over a circular hole cut in the top of the hover.

hurdle. A short, solid gate used when handling and herding hogs.

inbreeding. The mating together of closely related individuals to concentrate certain traits. As my late father-in-law said, "It can bring out the good and any of the bad, if it's there."

indexing. A score for a particular performance trait expressed as a number over or under the herd average. The average is often given a value of 100; better performers will score above 100, below-average performers score below it.

input. Data and information.

internal body dimension. Depth and width of a body cavity.

Jell-O middled. Soft and thick in the middle.

lactation. The period when the female is producing milk and nursing pigs. Producers now allow pigs to nurse from three to the traditional eight weeks. The shorter the lactation, the less draining it is on the sow, although three- to four-week weanings require special facilities for the weaned pigs and may affect the sow's recycling for breeding.

libido. The sexual drive that causes the male to seek out females, mount them correctly, and breed.

limit-feeding. The restricting of the amount of energy feedstuffs the animal receives daily. This is most often done to keep brood sows from becoming excessively fat and having farrowing difficulties. It can be done with growing hogs when feed efficiency and carcass trim can be improved by it, but days on feed will be extended.

line. A specific bloodline tracing back to a single individual or family of great merit.

line breeding. Repeat breeding to a single individual (parent to offspring), to concentrate that individual's genetics in a family.

litter. The offspring of a single farrowing. The record litter size is well in excess of 30; the national average is a fraction above 7 and has been for many years. Most producers hold that they need at least 6 pigs per litter to cover expenses, so 8 pigs per litter for gilts and 9 for sows are reasonable, reachable goals.

loineye. The area, in square inches, of lean meat contained within the loin. This is the most important cut economically — it is the pork chop — and the larger the loineye, the more it is worth to the consumer.

mad cow disease. A disease caused by an infectious particle of protein (a prion) most often seen in cattle. The disease causes reeling and unnatural behavior, hence the name. Also known as bovine spongiform encephalopathy (BSE).

market hog. Another term for a butcher hog. It weighs from 220 to 260 pounds and is usually 5 to 7 months of age when it goes to market.

meal. Finely ground feedstuffs or complete feeds. When ground too fine, some feedstuffs can present problems with palatability and ulcers.

meat type. Animals demonstrating trimness and good muscling patterns.

medicated early weaning (MEW). Similar to segregated early weaning. Pigs are weaned early and separated from pigs of other ages by moving them to an isolated facility to prevent the spread of disease, but sows also are immunized before farrowing.

metritis, mastitis, agalactia (MMA) complex. A disease complex causing fever, sudden hardening of the udder, and loss of milk flow.

middle of the road. This is a term of primary concern to the seedstock producer. It refers to an animal of good type, breed character, and performance but that is not of extreme type or a departure from industry thinking. It is a very balanced animal, a rather complete package. The producer of this type is referred to as a propagator rather than a breeder.

multiplier breeders. Those acquiring animals from an elite source, breeding them in greater numbers, and then marketing them to others.

mummies. Partially absorbed embryos that are sometimes seen with the placenta or at the time of an abortion. These are embryos that have died in late term; otherwise they would have been totally reabsorbed by the sow's system.

muscling. Visual indications of the hog's muscle pattern. It can be seen in ham and shoulder shape, the length that muscles are carried down the legs, and freedom of motion on top of the shoulders.

mycotoxins. The poisons produced by mold growth on grain and other feedstuffs. The younger the animal is, the more severely it is affected. Loss of baby pigs and abortions are the most common problems that result from mycotoxins.

natural immunity. The immunity that an animal or herd builds up over a period of time when exposed to disease-causing organisms. Sometimes it is transferred by survival of the fittest, sometimes via the blood during gestation, and sometimes during lactation.

natural ventilation. The ventilation that occurs naturally in cold housing.

navel cord. The cord by which the unborn pig was attached to the uterus and nourished. Sometimes during a farrowing, the cord may be severed before the pig clears the birth canal; the pig can suffocate, or the cord can twist about the pig's neck to produce the same result.

needle teeth. Also called wolf teeth, these are two large teeth on each side of the upper and lower jaw that are present at birth. If not clipped off — but not crushed — with sidecutters or the like, they can cut the sow's udder or other pigs, and infection will result. Some old-timers leave the small pig in a litter with its wolf teeth to give it a bit of an edge in the fight for a teat. Nursing order will be established in the first 48 hours of life or so, and once attached to a teat, a pig will return to it.

NIMBY. An acronym for "not in my backyard."

offal. Waste parts from a butchered animal.

omnivores. Creatures that eat both animal and vegetable foods.

open. A producer's term for a female of breeding age that isn't bred.

oral. Through or via the mouth as a means of treating an animal.

outcross. The use in the breeding program of an animal that has a completely different genetic background from any animals used previously. In purebred matings, animals with very distinct pedigrees can give their offspring a bit of what might be considered the equivalent of hybrid vigor.

oxytocin. A drug that can stimulate uterine contractions and milk letdown — and one that may well be overused in livestock.

parity. The condition or fact of having given birth.

pathogen. An agent that causes disease, such as a bacterium.

pelleting. Compressing feedstuffs into small, easily handled and fed pellets.

penile sheath. The fleshy sheath that houses and protects the penis. A large sheath can gather urine and semen or injure more easily and become a pocket of infection that reduces breeding performance. This trait seems to breed on and has shown up a great deal more with today's trend toward looser-sided hogs.

performance testing. Documenting growth rate, feed efficiency, and carcass traits. Animals in a group may be handled in groups for this.

phenotype. How the animal's genetic makeup manifests itself visually. The term covers traits such as conformation, soundness, and muscling, all somewhat apparent to the naked eye.

pickle cure. A brine cure.

pig. A very young pig.

piggy female. A female in the last half of gestation showing obvious signs of pregnancy.

pizzle. The penis and supporting tract.

placenta. The afterbirth that is also expelled at the time of farrowing.

premix. A vitamin-and-mineral mixture to be combined with grain and soybean oil meal to form a complete ration. This is quite often the least costly means of formulating a ration.

pressure treated. Runners, timbers, and structural lumber that have had a wood preservative applied to them under pressure. This forces the preservative deep into the wood to make it last longer.

prestarter. A feed very rich in protein and milk sugars offered as a first feed to baby pigs.

primal cuts. Large cuts of meat, often transported to butcher shops.

probiotics. Beneficial organisms similar to those already present in the gut that can be given to pigs and hogs to further improve gut activity and feed utilization. They may also be used to restore gut activity in sick and recuperating hogs.

production costs. All the costs required to provide a feeder pig or pound of pork. These should include taxes, interest, insurance, and dozens of other costs that can be overlooked in the planning stages of an enterprise.

pseudorabies. A swine disease of long standing that has grown in virility as other diseases were overcome.

purebred. The term describes animals that have been bred true for many generations and have traceable pedigrees.

race horse. Tall and leggy appearance.

ridgling. A male that retains one testicle within its body.

ringing. The crimping of a metal ring into the tip of the nose or across its end to keep the hog from rooting. When the pig tries to root, the ring causes it some discomfort.

rotation. This term can refer to the use of different breeds in an established order in a crossbreeding program. It can also refer to the movement of hogs to clean pastures or lots, to let previously used land lie fallow to break disease and parasite cycles.

rupture. A protrusion of a part of the intestine through an opening in the wall of muscle surrounding it. It's the same as a hernia.

saltpeter. Sodium nitrate; a curing product.

savage. To attack and injure others.

scours. The livestock producer's term for diarrhea.

scrotum. The saclike affair that carries the testicles. Semen is very heat sensitive — so much so that even the boar's body temperature can impair it — and it is the scrotum's purpose to both protect the testes and hold them close to or away from the body to protect the sperm as air temperatures warrant.

seedstock. The source of new swine stock.

segregated early weaning (SEW). In this practice, pigs are weaned early and moved to a nursery facility far from any other hogs as a method of breaking the disease cycle.

selective breeding. Putting together matings of animals strong in a particular trait and those weak in that trait to improve that trait in their offspring.

self-feeding. Allowing the animal to meet its own nutritional needs from a feeder that stores large amounts of feedstuffs and allows the animal unlimited access to them. This is also known as free-choice feeding.

service. An act of mounting and ejaculating is said to be a service by the boar.

set to the leg. This term describes front legs with good position at the corners of the body, cushioning in the feet, and a slope that supports and cushions the hog while it walks, stands to eat, and breeds.

settle. When a sow conceives, she is said to settle.

shed type. A type of housing with a slanting roof extending beyond both the high front wall and the lower rear wall. Doors and openings will be in the front wall, with some means of warm-weather ventilation in the rear.

shoat. A term used for a hog from 60 to 120 pounds.

silent heat. A female in estrus but demonstrating no visible signs.

sire. The male parent.

slaughter check. The examination of carcasses at the time of slaughter for clinical signs of illness. This type of examination is generally conducted by a veterinarian.

smallholding. A small, rural holding on which modest numbers of livestock are kept as part of a rural or simple lifestyle.

snare. A device for restraining hogs that fits around the snout.

sonoray. A device that uses ultrasound to measure lean muscling in a live hog.

sow. A female that has borne young.

soybean oil meal (SBOM). A protein supplement made from ground and heat-processed soybeans.

specific pathogen free (SPF). These are pigs taken from their dams by cesarean section and raised in isolation. The intention is to rear breeding animals free from certain health problems, such as sarcoptic mange. The approach was developed more than 30 years ago in Nebraska. These pigs are used to repopulate or reestablish herds that have had problems with disease. The SPF pigs are similar genetically.

split-sex feeding. Separating gilts and barrows, to be fed out in different groups.

spongy. Soft-middled and plump.

stag. A hog that has been castrated after reaching maturity.

standing heat. The female in estrus stands and eagerly accepts the male.

starter. A second-stage feed for the young pig, given from the time it is 14 days old until it weighs 40 to 50 pounds. These feeds have a crude protein content of 16 to 18 percent but are still rich in sugars and milk proteins.

stress. A strain or tension on the animal's well-being caused by weather, moving, weaning, changes in ration, castration, or other environmental or physical disruptions.

supplement. Feedstuffs added to grains to improve protein content and quality and to form a complete, balanced ration.

swine. A generic term roughly equal to (but a little fancier than) "hog."

tailender. A small, slow-growing animal in a large group.

tanbark. The fair or livestock show, termed thus because show-rings were once covered with spent bark, or tannin, a by-product of the process of tanning leather.

tankage. A protein supplement made from cooked and ground meat scraps.

terminal. A hog used as a cross to maximize growth and muscling, since the pigs produced are intended to be sold for butcher stock.

testosterone. A male hormone that is injected into a boar to stimulate sexual activity and libido. Ideally, it should be used only once or twice to get a slow-breeding boar started.

through the gut. This is the route preferred by many for treating a sick animal. Oral drugs, feed and water additives, and probiotic products all work via the gut to improve the animal's well-being in the way its systems work naturally.

top out. To have the highest sellers or performers in a market or competition.

top-dress. To place an additive or treatment on top of the hog's regular ration for consumption at the same time.

topline. The line that runs across the back of the animal; its top.

turbinates. Nose bones.

type. The physical appearance of the live animal, demonstrative of its growth and yield.

underline. The bottom line of the hog's body, on which are presented the penile sheath, udder segments, and teats. It is important that the teats be large and evenly spaced for best presentation and function.

unwinding. Dropping muscle control. The muscles grow slack, the animal slumps a bit, and the tail unwinds.

villi. Small, fingerlike tissues lining the gut that are key to digestion and absorption.

vulva. The external part of the female genitals. This is a good indication of reproductive-tract size and potential productivity.

walnut. A small, abscessed node.

wet sows. Sows that have recently weaned pigs and still have distended udders.

whole-hog sausage. Sausage made with all parts of the carcass.

wolf teeth. *See* needle teeth.

VOLUME CONVERSION

To convert	to	multiply
teaspoons	milliliters	teaspoons by 4.93
tablespoons	milliliters	tablespoons by 14.79
fluid ounces	milliliters	fluid ounces by 29.57
cups	milliliters	cups by 236.59
cups	liters	cups by 0.24
pints	milliliters	pints by 473.18
pints	liters	pints by 0.473
quarts	milliliters	quarts by 946.35
quarts	liters	quarts by 0.946
gallons	liters	gallons by 3.785

VOLUME EQUIVALENTS

US	Metric
1 teaspoon	5 milliliters
1 tablespoon	15 milliliters
¼ cup	60 milliliters
½ cup	120 milliliters
1 cup	240 milliliters
1¼ cups	300 milliliters
1½ cups	350 milliliters
2 cups	480 milliliters
2½ cups	600 milliliters
3 cups	700 milliliters

TEMPERATURE CONVERSION

To convert Fahrenheit to Celsius, subtract 32 from
Fahrenheit temperature, multiply by 5, then divide by 9.

Easy-to-Remember Equivalents

0°C = 32°F	30°C = 86°F
10°C = 50°F	40°C = 104°F
20°C = 68°F	Every 10°C = 18°F

WEIGHT CONVERSION

To convert	to	multiply
ounces	grams	ounces by 28.35
pounds	grams	pounds by 453.6
pounds	kilograms	pounds by 0.45

WEIGHT EQUIVALENTS

US	Metric
0.035 ounce	1 gram
¼ ounce	7 grams
½ ounce	14 grams
1 ounce	28 grams
1¼ ounces	35 grams
1½ ounces	43 grams
1¾ ounces	50 grams
2½ ounces	70 grams
3½ ounces	100 grams
4 ounces	113 grams
5 ounces	140 grams
8 ounces	227 grams
8¾ ounces	250 grams
10 ounces	284 grams
15 ounces	425 grams
16 ounces (1 pound)	454 grams

LENGTH / AREA CONVERSION

To convert	to	multiply by
inches	centimeters	2.54
feet	meters	0.31
yards	meters	0.91
miles	kilometers	1.61
square feet	square meters	0.09
acres	hectares	0.41

Page numbers in *italic* indicate illustrations or photographs.
Page numbers in **bold** indicate tables or charts.

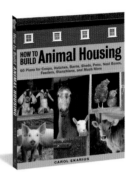

STOREY'S GUIDE TO RAISING

The Definitive Series for Essential Animal Husbandry Information

This best-selling series offers fledgling farmers and seasoned veterans alike what they most need to know to ensure both healthy livestock and profits. Each book includes information on selection, housing, space requirements, behavior, breeding and birthing, feeding, health concerns, and remedies for illnesses. They also cover business considerations and marketing products that come from the animals.

MORE THAN 2.2 MILLION COPIES SOLD!

THE COMPLETE STOREY'S GUIDE TO RAISING LIBRARY INCLUDES:

Beef Cattle
by Heather Smith Thomas

Chickens
by Gail Damerow

Dairy Goats
by Jerry Belanger and Sara Thomson Bredesen

Ducks
by Dave Holderread

Horses
by Heather Smith Thomas

Keeping Honey Bees
by Malcolm T. Sanford and Richard E. Bonney

Meat Goats
by Maggie Sayer

Miniature Livestock
by Sue Weaver

Pigs
by Kelly Klober

Poultry
by Glenn Drowns

Rabbits
by Bob Bennett

Sheep
by Paula Simmons and Carol Ekarius

Training Horses
by Heather Smith Thomas

Turkeys by Don Schrider